BOB MILLER'S GEOMETRY FOR THE CLUELESS

GEOMETRY

OTHER TITLES IN BOB MILLER'S CLUELESS SERIES

BOB MILLER'S GEOMETRY FOR THE CLUELESS

GEOMETRY

Robert Miller

Mathematics Department
City College of New York

McGraw-Hill

New York   San Francisco   Washington, D.C.   Auckland   Bogotá
Caracas   Lisbon   London   Madrid   Mexico City   Milan
Montreal   New Delhi   San Juan   Singapore
Sydney   Tokyo   Toronto

Library of Congress Cataloging-in-Publication Data

Miller, Robert.
   [Geometry for the clueless]
   Bob Miller's geometry for the clueless : geometry / Robert Miller.
     p.   cm. — (Bob Miller's clueless series)
   Includes index.
   ISBN 0-07-136109-X
   1. Geometry.   I. Title: Geometry for the clueless.   II. Title: Geometry.
  III. Title.
  QA445 .M512 2000
  516—dc21   00-055404

Sponsoring Editor: Barbara Gilson
Production Supervisor: Charles Annis
Editing Supervisor: Maureen B. Walker
Compositor: North Market Street Graphics
Photo: Eric Miller

McGraw-Hill

A Division of The McGraw-Hill Companies

To my dearest Marlene, I dedicate this book and everything else I ever do. I love you very, very much.

TO THE STUDENT

This book was written for you: not your teacher, not your neighbor, not for anyone but you.

However, as much as I hate to admit it, I am not perfect. If you find anything that is unclear or should be added to the book, please let me know, but address your comments c/o McGraw-Hill, Schaum Editorial Director, Two Penn Plaza, Eleventh Floor, New York, New York 10121. Please enclose a self-addressed stamped envelope. Be patient. I will answer.

If your basic algebra is a little shaky, my *Algebra for the Clueless* will help. For more advanced stuff, my *Precalc with Trig for the Clueless* and *Calc I, II, and III for the Clueless* will help.

For the SAT, my *SAT Math for the Clueless* will be just what you need.

Now enjoy this book and learn!!!

CONTENTS

WHAT IS GEOMETRY? WHY SHOULD I TAKE IT?

The geometry we study, called Euclidean geometry, was written down about 2300 years ago in ancient Greece. It is an example of a *logic system*. At the start of every logic system are *undefined words*. Although Euclid had many more undefined words, we have narrowed the list down to five but will use six for convenience. Here they are:

1. Point

2. Line (infinite straight)

3. Plane (this one is usually undefined for convenience)

4. On (yes indeed! On!!!)

5. Between

6. Congruent (identical in every way)

Notice what happens if we try to define the word "on" as "not off." We are forced to define the word "off," which can only be defined as "not on." On and on and on. You see the problem. So "on" is undefined.

Next are the *defined words*. There are many. Let us list three. They will be repeated.

1. Line segment: All *points on* a *line between* two *points.*

2. Ray: All *points on* a *line on* one side of a *point on* that *line.*

3. Angle: two *rays* with a common end*point.*

Notice all the undefined words in these three definitions. The last definition also uses a previously defined word.

Next there are *axioms* or *postulates,* laws taken to be true without proof. When I went to school, we used axioms for algebraic laws and postulates for geometric laws. I guess I'll do that, but *every* book is different. You must check your book.

One postulate is all radii of a circle are equal or congruent, depending on the book (you can't measure them all).

The last is a *theorem,* a law that is proven. One example is the sum (of the measures) of the angles of a triangle is 180 degrees, a fact we will need for this course and the SAT.

Now that you have an idea what geometry is, you might ask, "Why should I take geometry?"

This is a very good question. Although it is necessary for students of math, it is just as important if not more for nonmath students. Let me give you two examples.

A geometric proof is like a history essay. First you are given some facts and a question to answer. Then in a series of logical steps, prove the answer, exactly a geometric proof without pictures.

Somewhere between your high school junior year and sophomore college year, you may read Plato's *Republic.* Plato, a Greek philosopher, tried to prove what a perfect society would be. One of his undefined words was *good.* We all have an idea what a *good* person is. Plato then adds defined words, axioms (postulates), and theorems, and he proves what a perfect society would be. Again, this is geometry without pictures. There are many more examples.

Finally you may ask, "Do I have a good book?"

For a book to be good, it must be clear and logical, but not necessarily easy. Learning how to think can be compared to body building. Both are hard, but both are very satisfying after you have accomplished your purpose. Also, in general, the more nonmemorized proof days you spend, the better the course. But we'll get very specific in the book.

You may notice that I have two definitions for some words. Also some theorems will have two proofs, an old proof (before 1970) and a new proof. The problem with the newer proofs is that too much precision is sacrificed for understanding and logical thinking. Let me give you an example:

> The page on the right is from a book of a 6-year-old child.

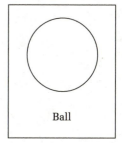

Ball

> The instruction says color the ball red.

> The child picks up a red crayon and colors the ball red.

> Buuuuuuut, the direction is not precise.

> To the right is not a ball but a picture of a ball.

> The precise direction is color the *picture* of the ball red.

> However the child now does not know whether to color the ball or the whole page: more precision, less understanding.

> I've tried to write it so you'll understand either way!

O K A Y !!!!!!!! Let's get going!!!!!!!!!

BOB MILLER'S GEOMETRY FOR THE CLUELESS

GEOMETRY

CHAPTER 1

THE BASICS: UNDEFINED WORDS, SOME DEFINED WORDS, AXIOMS, AND POSTULATES

UNDEFINED WORDS

At the start of geometry there are certain words that you cannot define, but we all know what they are. Let's look at them, explain them a little, and give their notation.

The first is *point.* Points are indicated with one capital letter. In the picture, A, B, and C are points.

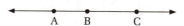

The second is *line.* By a line we mean infinite straight line. In the picture the line is indicated by $\overleftrightarrow{AB}$. But we could have also used $\overleftrightarrow{AC}$, $\overleftrightarrow{BC}$, $\overleftrightarrow{BA}$, $\overleftrightarrow{CA}$, or $\overleftrightarrow{CB}$. All are correct.

The third is *on.* The point is *on* the line. Notice if you try to define on as not off, then you have to define off. You go in circles.

The fourth is *between.* On the line B is *between* A and C. Some books define between. Later we will give the definition.

The last is *congruent,* identical in every way. Its symbol is ≅.

These words are not hard, but like all definitions, you must know them verrry, verrry well.

THE BEGINNING OF DEFINED WORDS

We are now starting to build up geometry. Let us define words that most of us know but that we need to know more precisely.

Line Segment   All points on a line between two points.

The notation for our line segment is either AB or $\overline{AB}$. We will use AB. Notice how many undefined words we used.

Ray   All points on one side of a point on a line.

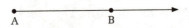

The notation for a ray is $\overrightarrow{AB}$. This means all points starting from A, called the *endpoint,* through B. The ray $\overrightarrow{BA}$ is the ray with endpoint B through A.

Angle   Two rays or two line segments with a common endpoint.

   The common endpoint is called the *vertex* of the angle. Each angle pictured is read ∠ABC or ∠CBA. The vertex is allllways in the middle. The angle can also be written ∠B, by just naming the vertex only.

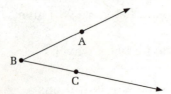

However in this picture you cannot use only one letter since there are three angles with vertex B. Do you see them? They are ∠ABC, ∠CBD, annnd ∠ABD.

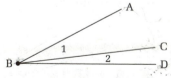

If I were very lazy, and I often am, I could have labeled ∠ABC as ∠1 and ∠CBD as ∠2 as in the picture. But do so only if your teacher allows it.

∠ABC and ∠CBD are called *adjacent angles* because they are two angles with a common vertex and common side (BC) between. ∠ABC and ∠ABD are *NOT* adjacent angles because the common side AB is not, not, not between the angles.

Many times in this book I will give you two definitions or two explanations. At the beginning, it will be because there are mainly two definitions used in many books. However later, when we have other definitions or start proofs, I will give two proofs. One of them will be from when I took geometry. Since sometimes my kids call me ancient, I will call them ancient proofs. I put those in because you will understand them better. The other will be called newer, one that might be in your book.

Warning!!!!!! Warning!!!!!! Every geometry book is different. You must use the statements that are in your book. This book is written for you to understand but may be slightly different than your book. *Warning!!!!!! Warning!!!!!!!!!*

ANGLES AND LINE SEGMENTS THE SAME

Ancient Explanation   ∠A = ∠B. If you were to angle A on top of angle B, they would appear to be the same angle. Although everyone understood this definition, according to the new math of the 1960s and 1970s (the worst thing to hit math until the new texts of the 1990s) this explanation is not technically correct.

Newer Explanation   Physical objects, the angles themselves can only be *congruent*. $\angle A \cong \angle B$. However if both angles are 27° their *measures* are equal. We write $m\angle A = m\angle B$. We then make the following definition.

DEFINITION OF CONGRUENT ANGLES
$\angle A \cong \angle B$ if $m\angle A = m\angle B$. Similarly with line segments.

Ancient Explanation $AB = CD$. If you put AB on top of CD, they would be exactly the same length, and you would see only one segment.

Newer Explanation $AB \cong CD$ if $mAB = mCD$. Fortunately most books don't do this. This is an inconsistency for which we are very happy.

SUPER IMPORTANT DEFINITIONS AND A FEW POSTULATES

Whenever anyone has trouble with geometric proofs, I always check this section first. You must know this section perfectly.

Postulate I   Once around a circle is 360° (360 degrees). (Easy.)

Postulate 2a   Line segments (ancient version): The whole is equal to the sum of its parts $AB + BC = AC$. In some books (newer version): Definition of between is

AB + BC = AC (AB + BC ≅ AC). Also, in some newer books, this is called segment addition postulate. Whew!!!

Postulate 2b   Angles (ancient version): The whole is equal to the sum of its parts. ∠ABC + ∠CBD = ∠ABD. And (newer version): Angle addition postulate m∠ABC + m∠CBD = m∠ACD.

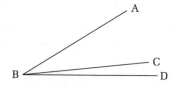

DEFINITION OF MIDPOINT
M is the midpoint of AB if (ancient) AM = MB or (newer) AM ≅ MB. Alternate definition: (ancient) AM or MB = 1/2 AB or (newer) AM or MB ≅ 1/2 AB.

A——————•——————B
        M

DEFINITION LINE SEGMENT BISECTOR
CD bisects AB at E if (ancient) AE = EB or (newer) AE ≅ EB.

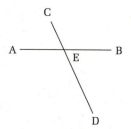

Angle Bisector   BD bisects angle ABC if ∠ABD = ∠DBC (∠ABD ≅ ∠DBC).

Perpendicular: ⊥ Two lines (line segments) are perpendicular if they form right angles, angles of 90°. On the picture if AB ⊥ CD, then rt ∠ABC = (≅)rt

---

NOTE

When you bisect an orange, you get two equal (congruent) orange halves. When you bisect an angle, you get two equal (congruent) angles. It is NOT NOT NOT like a wishbone.

In many geometry books, the definition of perpendicular is two lines or line segments that form equal (congruent) adjacent right angles. *You must check your book, for I have seen many different definitions for perpendicular, more than any other word.*

∠ABD since perpendiculars form equal (congruent) right angles. In proofs, when we see perpendicular, we will write this statement out, usually leaving out rt (right).

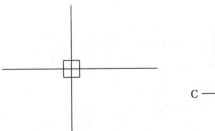 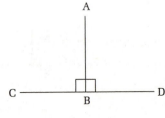

Note the spelling of complementary (complE . . .). Complimentary (complI . . .) is me telling you how good looking and smart you are (a compliment).

Complementary Angles   Any two angles whose sum is 90 degrees (sum of the measure of two angles is 90 degrees). In symbols ∠1 + ∠2 = 90° (m∠1 + m∠2 = 90°). The most common picture is below.

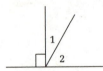

Supplementary Angles   Any two angles whose sum is 180 degrees. In symbols ∠1 + ∠2 = 180° (m∠1 + m∠2 = 180°). The usual picture is below. Sometimes supplementary angles as pictured is called a *linear pair.*

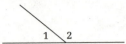

Vertical Angles   As pictured, angles 1 and 2 and angles 3 and 4 are vertical angles.

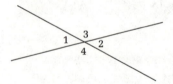

Theorem (a statement to be proven): Vertical angles are equal (congruent). In symbols ∠1 = ∠2 (∠1 ≅ ∠2). Also ∠3 = ∠4 (∠3 ≅ ∠4).

IMPORTANT IMPORTANT IMPORTANT IMPORTANT IMPORTANT IMPORTANT

If you do not know the definitions of midpoint, angle bisector, segment bisector, complementary, supplementary, and vertical angles, memorize and know them absolutely perfectly, you will not be able to do the proofs.

I know more lists are boring, at least to some of you, but we need more facts.

This theorem is in your book, so I won't prove it. Anyway, we are not ready for proofs just yet. We need some more axioms, postulates, and definitions.

SOME AXIOMS

Axiom 1: Addition If equals (congruents) are added to equals (congruents), their sums are equal (congruent).

x = 5	m∠A = 32°	∠C ≅ ∠D
y = 2	m∠B = 18°	∠F ≅ ∠G
x + y = 7	m∠A + m∠B = 50°	∠C + ∠F ≅ ∠D + ∠G

Axiom 2: Subtraction If equals (congruents) are subtracted from equals (congruents), their differences are equal (congruent).

3x + 5y = 20	m∠1 = 57°	a ≅ b
2x + 5y = 11	m∠2 = 23°	c ≅ d
x = 9	m∠1 − m∠2 = 34°	a − c ≅ b − d

Axiom 3: Multiplication If equals (congruents) are multiplied by equals (congruents), their products are equal (congruent).

$3x = 7$	$1/4\ AB = 5$	$\frac{1}{2}m\angle F = 27°$	$a \cong b$
$12x = 28$	$AB = 20$	$m\angle F = 54°$	$2a \cong 2b$

Axiom 4: Division   If equals (congruents) are divided by nonzero equals (congruents), their quotients are equal (congruent).

$12x = 36$ $\qquad AC \cong FH$

$\quad x = 3$ $\qquad 1/3\ AC \cong 1/3\ FH$

Axiom 5: Reflective Law   (In some books, it is called the identity.) A quantity is equal to (congruent to) itself.

$a = a \qquad x + y = x + y \qquad m\angle C = m\angle C \qquad AB \cong AB$

Axiom 6: The Symmetric Law   (This is usually used to switch sides in an equation.) If $a = b$, then $b = a$. (If $a \cong b$, then $b \cong a$.) If $a + b = y$, then $y = a + b$. If $CD \cong AB$, then $AB \cong CD$.

Axiom 7: Transitivity   (My favorite.) If $a = b$ and $b = c$, then $a = c$. (If $a \cong b$ aaaand $b \cong c$, theeeennnnn $a \cong c$.) If $x = y$ and $y = 7$, then $x = 7$. If $AB \cong BC$ and $BC \cong CD$, then $AB \cong CD$.

NOTE
There are other transitive laws If $a < b$ and $b < c$, then $a < c$. The same is true for $\leq, >, \geq$.

Axiom 8: Substitution   (Technically it can be proven, but usually is not. So we call it an axiom.)

  A. Quantities equal to (congruent to) the same quantity are equal to (congruent to) each other. If $a = x$ and $a = y$, then $x = y$. If $AB \cong XY$ and $AB \cong YZ$, then $XY \cong YZ$.

B. Quantities equal (congruent) to equal (congruent) quantities are equal (congruent) to each other. If a = x, b = y, and a = b, then x = y. If $\angle A \cong \angle X$, $\angle B \cong Y$, and $\angle A \cong \angle B$, then $\angle X \cong \angle Y$.

C. If a equals (is congruent to) b, then a may be substituted for b in any algebraic or geometric true statement. a = b. If a + x = y, then b + x = y. AB $\cong$ CD. If MQ $\cong$ 4 AB, then MQ $\cong$ 4 CD.

There are more. Sorry. I really really am. But we must.

THE LAST DEFINITIONS (OF TYPES OF ANGLES AND PLANE) AND POSTULATES BEFORE THE PROOFS

Acute angle: Any positive angle that is less than 90°.

Tight angle: That can't be correct. Ah! *Right angle:* A 90° angle.

Obtuse angle: An angle of more than 90° but less than 180°.

Straight angle: An angle of 180°.

Plane: In some books this is an undefined word. (It is an infinite flat surface.) In some books, it is defined as the set of all lines that can be drawn through two intersecting lines.

See, if you draw enough lines, you see the plane start to form.

360°

Postulate 1 (again) Once around a circle is 360°.

Postulate 2 (for the first time!) One and only one line can be drawn between two points.

Postulate 3 Two lines cannot intersect at more than one point. (If they intersected at two points, by postulate 2 there would be only one line.)

Postulate 4 A line segment is the shortest distance between two points on a plane.

Postulate 5 Any geometric shape can be moved without changing its shape.

Postulate 6 A line segment has only one midpoint.

Postulate 7 An angle can have one and only one bisector.

Postulate 8 The perpendicular is the shortest line segment between a point and a line.

Postulate 9 The sum (of the measures) of the angles on one side of a line is 180°.

Postulate 10 Two adjacent angles are supplementary if their exterior sides are in a straight line.

I know most of these statements are obvious, but since we are not proving them or they cannot be proved, we have to list those postulates.

Now on to the beginnings of proofs???!!!!!!!!!!!!

CHAPTER 2

THE BEGINNINGS OF PROOFS

This is the reason you probably bought this book, so I'd better make this chapter great. Seriously, I hope you think all the chapters are great. I'm sure working hard enough for that to happen.

NOTE 1

Some of you probably started to read this book from here. You really should read *everything* that came before. It is verrry important.

NOTE 2

How good and valuable your geometry book is depends on how many weeks you spend on doing original proofs. If it is half your year or more you have a very good book and course. If you spend only a few weeks on proofs, you have a bad book. There is one crummy geometry book with only one chapter on proofs. Really bad!!!!!

Formal proofs are meant to train you logically and are very exact. Here are some beginning steps. More will follow as the proofs get longer. Notice I didn't say harder. Once you get the hang of these, it won't matter how long.

1. In the upper left, draw the picture, a larrrge picture, with a ruler, even if the teacher doesn't. If the drawing is too sloppy or too small, you may not see what you should see!!!!

2. In the upper right, put the hypothesis (the facts you are given). In some theorems at the beginning, there might not be a hypothesis.

3. Under the given, write the conclusion, what you are trying to prove.

4. Draw a horizontal line under everything. Draw a vertical line about $\frac{1}{3}$ of the way on this line as long as you think the proof will take.

5. On the left are your statements you make.

6. On the right are your reasons the statements are true: undefined words, defined words, axioms, postulates, and previous theorems.

7. The last line under the statements should be exactly what you wanted to prove.

Last, we will do many proofs two ways: the ancient way (for more clarity) and a more recent way. Sometimes we will do the proofs together (to be brief). If we do, we will do the ancient way and the newer way will be in (parentheses) as we did before.

Okay, okay!!!!!! Enough talking. Boy, I talk too much!!!! Let's do the first proof.

Theorem 1   Given line segment AD with points B and C between A and D. If AB = CD (AB $\cong$ CD), then AC = BD (AC $\cong$ BD).

Picture   A   B   C   D

What you should see: AB and CD are the same. Make sure when you draw the picture that they are the same!!! How do I get AC? I add BC to AC. How do I get BD? I add BC to CD. AHA!!! I am adding equals to equals.

It will take you a while to learn the style. But you can get it, and you *will* get it!!!!!!!!

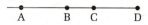

A   B   C   D

Given: 1. ABCD is a line segment   2. AB = CD (AB ≅ CD)

Prove: AC = BD (AC ≅ BD)

Statement	Reason
1. ABCD is a line segment	1. Given (1 and 2 are true because we said so).
2. AB = CD (AB ≅ CD)	2. Given (we now add BC to both).
3. BC = BC (BC ≅ BC)	3. Reflexive law or identity law (depending on the book).
4. AB + BC = (≅) BC + CD	4. If = s(≅s) are added to =s(≅), their sums are =(≅).
5. AB + BC =(≅)AC BC + CD =(≅)BD	5. The whole is equal to the sum of its parts or definition of between or segment addition postulate.
6. AC = BD (AC ≅ BD)	6. Substitution (step 5 into step 4).

The difficult part of geometry is the discipline of putting everything you do down!!!!!. For most people, this is verrry DIFFICULT in the beginning. Reread the proofs. Write them down. You will get it. It is just a matter of time and sticking with it. Let's do a similar one.

Theorem 2   ABCD is a line segment. AC = BD(AC ≅ BD). Prove AB = CD (AC ≅ BD). The picture is the same as the above. What you should see: Going from larger line segments to smaller line segments, you must subtract. However, the steps are in a slightly different order.

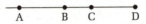

Given: ABCD a line segment
         AC = BD (AC ≅ BD)

Prove: AB = CD (AB ≅ BD)

Statement	Reason
1. ABCD a line segment	1. Given.
2. AC = BD (AC ≅ BD)	2. Given.
3. AC = AB + BC (AC ≅ AB + BC) BD = BC + CD (BD ≅ BC + CD)	3. The whole equals the sum of its parts or definition of between or segment addition postulate.
4. AB + BC = BC + CD (AB + BC ≅ BC + CD)	4. Substitution (3 into 2).
5. BC = BC (BC ≅ BC)	5. Reflexive or identity law.
6. AB = CD (AB ≅ CD)	6. If = s(≅) are subtracted from =s(≅), their differences are =(≅).

NOTE
Most teachers may allow you to take short cuts when you write the reasons, and I probably will too just to save space. But write out the whole reason. My high school teacher Miss Griswold made us write all the reasons out. I was not a natural at geometry, but she made me very good. She was the best math teacher I

ever had. Remember I have had a lot of math teachers!!!!!!!!!

Let's do a similar proof with angles.

Theorem 3

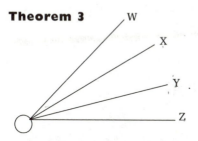

Given: ∠WOX = ∠YOZ (∠WOX ≅ ∠YOZ)

Prove: ∠WOY = ∠XOZ (∠WOY ≅ ∠XOZ)

What you should see: This theorem is very similar to theorem 1, only with angles.

Statement	Reason
1. ∠WOX = (≅) ∠YOZ	1. Given.
2. ∠XOY = (≅) ∠XOY	2. Reflexive (identity).
3. ∠WOX + ∠XOY = (≅)∠XOY + ∠YOZ	3. If =s(≅s) are added to =s(≅), their sums are =(≅).
4. ∠WOX + ∠XOY = (≅)∠WOY	4. The whole =s the sum of its parts (∠addition postulate)
5. ∠XOY + ∠YOZ = (≅)∠XOZ	5. Same as 4.
6. ∠WOY = (≅)∠XOZ	6. Substitution (4 and 5 into 3).

Once you get the flow of the proofs, they will go much easier.

You might try to do the next theorem on your own. It is similar to theorem 2 but with angles like those in theorem 3.

Theorem 4   Same picture. Given ∠WOY = (≅)∠XOZ.
Prove ∠WOX = (≅)∠YOZ.

The next proof shows what happens when the new math was put into effect. The proofs are different enough so we will do it twice.

Theorem 5a Supplements (sups) of equal angles (or the same angle) are equal.

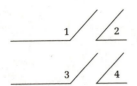

$\angle 1 = \angle 3$

Given: $\angle 1$ is the supplement of $\angle 2$

$\angle 3$ is the supplement of $\angle 4$

Prove: $\angle 2 = \angle 4$

Statement	Reason
1. $\angle 1$ is the sup of $\angle 2$	1. Given.
2. $\angle 1 + \angle 2 = 180°$	2. Definition of supplementary.
3. $\angle 3$ is the sup of $\angle 4$	3. Given.
4. $\angle 3 + \angle 4 = 180°$	4. Definition of supplementary.
5. $\angle 1 + \angle 2 = \angle 3 + \angle 4$	5. Substitution of 4 into 2.
6. $\angle 1 = \angle 3$	6. Given.
7. $\angle 2 = \angle 4$	7. If =s are subtracted from =s, their differences are =.

Theorem 5b

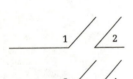

$\angle 1 \cong \angle 3$

Given: $\angle 1$ is the sup of $\angle 2$

$\angle 3$ is the sup of $\angle 4$

Prove: $\angle 2 \cong \angle 4$

Statement	Reason
1. $\angle 1$ is the sup of $\angle 2$	1. Given.
2. $m\angle 1 + m\angle 2 = 180°$	2. Definition of sups.
3. $\angle 3$ is the sup of $\angle 4$	3. Reason 1.
4. $m\angle 3 + m\angle 4 = 180°$	4. Reason 2.
5. $m\angle 1 + m\angle 2 = m\angle 3 + m\angle 4$	5. Substitution of 4 into 2.
6. $\angle 1 \cong \angle 3$	6. Reason 1.
7. $m\angle 1 = m\angle 3$	7. Definition of congruent angles.
8. $m\angle 2 = m\angle 4$	8. Subtraction (step 5 minus step 7).
9. $\angle 2 \cong \angle 4$	9. Reason 7.

There are a number of things to notice.

NOTE 1
Notice the way the proof flows. *Everything* has to be explained carefully (and written down) and logically. It is the way a lawyer builds a case to convict someone of a crime. The jury knows nothing. Everything is built up step by step until guilt is proven. The same holds for geometric proofs.

NOTE 2
Notice the given is not put down in one step and not necessarily in the order given. This is so the story is totally logical, from beginning to end.

NOTE 3
Notice the second way we proved this theorem we added to steps which made the proof a little longer and a little more difficult to understand and follow. Because of the new math distinguishing unnecessarily between physical angles and measures of angles,

proofs have been made unnecessarily longer and pickier. But most modern books do it this way, so we are forced to do it.

NOTE 4

You should look at each theorem several times. Once you see the flow of the theorem, you will not have much trouble, if any.

Theorem 6 Complements of equal (congruent) angles or the same angle are equal (congruent). The proof of this theorem is exactly the same as the proof of theorem 5 except . . . . the word supplement is replaced by the word complement.

aaaaaand 180° is replaced by 90°. The new picture is at the right. You should try the proof.

Let's do a couple of word problems. As you will see, geometry word problems are really easy. In fact the two here are probably the hardest possible, and they are easy, if your basic algebra is OK.

EXAMPLE 1

An angle is 4 times its complement. Find the angle. Let x = angle. Thennnn $90 - x$ is its complement. The equation is

$$x = 4(90 - x) \qquad x = 360 - 4x \qquad 5x = 360$$
$$x = 72 \text{ degrees} \qquad 90 - x = 18 \text{ degrees}$$

Notice 4 times 18 is 72.

Let's try a slightly more complicated one, about as hard as you would get in all of geometry.

EXAMPLE 2—

Four times an angle is the supplement of its complement. Find the angle. x = angle. $90 - x$ is the complement of the angle. $180 - (90 - x)$ is the supplement of the complement. Soooo

$$4x = 180 - (90 - x) \qquad 4x = 90 + x \qquad 3x = 90$$
$$x = 30 \text{ degrees}$$

Let's go on.

We still have some more introduction to do, believe it or not. We need to do a little formal logic, a very important part of geometry.

Statement

If John eats glicks, then Mary sees flegs. As we said before the *given* follows "if," and the *conclusion* follows "then."

If we reverse the given and the conclusion, we form the *converse.*
   If Mary sees flegs, then John eats glicks.
If in the original statement we add the word *not* to each part, we have the *inverse.*
   If John does not eat glicks, then Mary does not see flegs.

If we add the word not to each part of the converse, we get the *contrapositive.*
   If Mary does not see flegs, then John does not eat glicks.

In logic, the statement and the contrapositive are either both true or both false together. The inverse and the converse are either both true or both false together. A geometric example, which we will do in greater detail later, should help.

Statement: A square is a rectangle. True. (It is a rectangle with equal (congruent) sides.)

Converse: A rectangle is a square. False. (The sides may not be the same.)

Inverse: If it is not a square, then it is not a rectangle. False. (It might not be a square because the sides are not the same.)

Contrapositive: If it is not a rectangle, then it is not a square. True.

Last, a first-class geometry course has some *indirect proofs.* Up to this time, we have done *direct proofs.* Let's define an indirect proof and give one example. We will come back a few times during the book.

Indirect Proof   We list all choices. We show all but one are false. The remaining one must be true.

DEFINITION: TRICHOTOMY LAW

In comparing two numbers exactly one is true: $a < b$, $a = b$, $a > b$. Translation: If we compare two numbers, the first is either less than the second, equal the second, or greater than the second.

If we can prove $a > b$ is false and we can prove $a < b$ is false, the only conclusion is that $a = b$.

CHAPTER 3

PARALLEL LINES, FOREVER TOGETHER

This is a topic that most students like a whole lot. Let us define.

DEFINITION: PARALLEL LINES

Two lines *in a plane* that never meet. In space you can have nonparallel lines that never meet. They are called *skew* lines. Cross your arms without letting them touch each other. Your arms are skew. Let us list some facts about parallel lines.

Postulate 1   In a plane, lines are either parallel or they meet.

Postulate 2   Very, very important: Through a point in a plane not on a line, one and only one line can be drawn parallel to that line.

Fact 1   Parallelism *is not* reflexive: line a ∥ line a is false. A line cannot be parallel to itself.

Fact 2   Parallelism *is* symmetric: if a ∥ b, then b ∥ a.

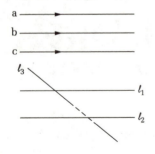

Fact 3   Parallelism *is* transitive: If a ∥ b and b ∥ c, then a ∥ c.

Fact 4   In a plane, if a line intersects one of two parallel lines, it intersects the other.

Facts 2–4 are small theorems and are not proved here. We need a few definitions for parallel lines before we start.

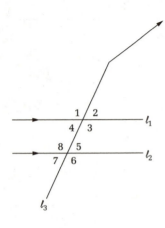

l_3 is called a transversal: a line which cuts two or more other lines. We will use transversals to cut parallel lines.

l_1 and l_2 are parallel in this picture.

Angles 1 and 8 are called *corresponding angles*. In the two groups of angles they are in the upper left. So are 2 and 5 (upper right), 3 and 6 (lower right), and 4 and 7 (lower left).

Angles 3 and 8 are called *alternate interior angles* because they are on the opposite side of the transversal and between the lines. So are 4 and 5.

Angles 2 and 7 and 1 and 6 are called *alternate exterior angles*.

Angles 3 and 5 and 4 and 8 are *interior angles on the same side of the transversal*.

Angles 2 and 6 and 1 and 7 are *exterior angles on the same side of the transversal*.

The following theorems are very important.

Theorem 1   If two parallel lines are cut by a transversal, alternate interior (exterior) angles are equal (congruent).

Theorem 2   If two parallel lines are cut by a transversal, corresponding angles are equal (congruent).

Theorem 3   If two parallel lines are cut by a transversal, interior (exterior) angles on the same side of the transversal are supplementary.

As important, all the converses to these theorems are true, which, in general, does not happen.

Converse to Theorem 1 If two lines, which are cut by a transversal, have alternate interior (exterior) angles equal (congruent), then the lines are parallel.

Converse to Theorem 2 If two lines, which are cut by a transversal, have corresponding angles equal (congruent), the lines are parallel.

Converse to Theorem 3 If two lines, which are cut by a transversal, have interior (exterior) angles on the same side of the transversal supplementary, then the lines are parallel.

NOTE 1

A. To find alternate interior angles, look for the *letter Z or backwards Z.*

B. For corresponding angles, look for the *letter F or backwards F or upside down F.*

C. For vertical angles, look for the *letter X.*

D. For interior angles on the same side of the transversal, look for a *square C (or backwards square C).*

NOTE 2

The SAT likes these questions. But you don't need to know them as well. You need to know that angles that look equal are equal. If they are not equal, they add to 180°.

NOTE 2

These theorems are proved by indirect proofs. We will do one of them later although I'm not sure which one.

NOTE 3

In this picture, we have vertical angles, which are the same: 1 and 3, 2 and 4, 6 and 8, 5 and 7.

NOTE 4

Teachers like to put numerical problems on parallel lines on tests. Let's do some now and save a few proofs for later.

The following picture below is for the next three examples.

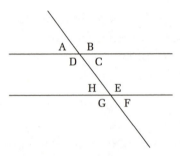

EXAMPLE I—

C is 4x − 2. ∠H is 2x + 24. Find x.

C and H are alternate interior angles. See the Z!!!

Sooooooo

$$4x - 2 = 2x + 24$$

$$ +2 +2$$

$$4x = 2x + 26$$

$$-2x -2x$$

$$2x = 26$$

$$x = 13$$

If we wanted the angles C is 4(13) − 2 = 50 (degrees); H = 2(13) + 24 = 50, as it must be.

EXAMPLE 2—

$\angle$C is x + 5. $\angle$E is 4(x + 10). Find x.

C and E are interior angles on the same side of the transversal. They are sups.

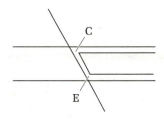

See the "C."

$x + 5 + 4(x + 10) = 180$

$x + 5 + 4x + 40 = 180$

$5x + 45 = 180$

$5x = 135$

$x = 27$

The angles are x + 5 = 32; 4(x + 10) = 4(27 + 10) = 4(37) = 148; 32 + 148 = 180, as it must.

EXAMPLE 3—

$\angle$B is 7x − 2. $\angle$D is x + 88. Find x. See the big X. B and D are vertical angles.

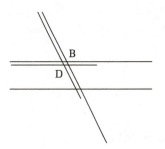

$7x − 2 = x + 88$

$x = 15$

Both vertical angles are 103 degrees.

CHAPTER 4

MOSTLY TRIANGLES

This section is the beginning of the largest and most important part of the book, if you have a good book. We will need some more definitions and examples.

DEFINITION: POLYGON

A closed connected figure. The figure on the right has five sides. It is called a *pentagon.* Each point is called a *vertex (plural is vertices).*

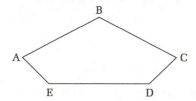

DEFINITION: QUADRILATERAL

A four-sided polygon. Examples are square, rectangle, trapezoid, rhombus, parallelogram. . . . We will look at these a lot later.

DEFINITION: TRIANGLE

A three-sided polygon. Triangles can be classified in two different ways: by sides or by angles.

Scalene triangle: a triangle with three unequal (non-congruent) sides and three unequal (noncongruent) angles.

The very popular *isosceles triangle:* A triangle with two equal (congruent) sides.

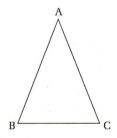

1. AB, AC *legs:* the equal (congruent) sides.

2. BC *base:* usually not the same length as the legs, but might be. The base also doesn't have to be at the bottom, although it usually is.

3. ∠B and ∠C, the *base angles:* always opposite the legs; always equal (congruent).

4. ∠A, the *vertex angle:* always opposite the base; usually different from the other angles but not always; usually at the top of the triangle but not always.

Equilateral triangle: A triangle with all sides the same; it is also *equiangular;* each angle is 60° since, as we will soon show, the angles of a triangle total 180°.

Acute triangle: A triangle with all three angles less than 90°.

The most important one *right triangle:* a triangle with 1 right angle.

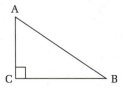

1. ∠C is the *right angle,* denoted as in the picture.

2. ∠A and ∠B are acute angles. They usually are not the same.

3. AB is the *hypotenuse,* always opposite the right angle.

4. It is always the largest side since the largest side lies opposite the largest angle, and the largest angle lies opposite the largest side.

5. AC and BC are the legs which lie opposite the acute angles. They usually are unequal.

Obtuse triangle: a stupid triangle; just kidding; a triangle with one obtuse angle.

Let's give examples of combinations of triangles.

Scalene and acute: angles are 50°, 60°, and 70°. The sides would also all be different.

Scalene and right: angles 30°, 60°, and 90°. We will see this one later.

Scalene and obtuse: angles 20°, 50°, and 110°.

Isosceles and acute: angles 50°, 50°, and 80°.

Isosceles and right: 45°, 45°, and 90°. We will see this one later, also.

Isosceles and obtuse: 20°, 20°, and 140°.

Buuut there is only one *equilateral* and *equiangular* triangle: 60°, 60°, and 60°.

Sorry, but we need a few more definitions. Remember this book has only the most essential definitions. At least, one is a review.

Altitude: AD   A line segment drawn from a vertex perpendicular to the opposite side and extended if necessary (as in the second figure).

> **NOTE**
>
> A triangle can't have two right angles since the third angle would have to be zero degrees.

> **NOTE**
>
> A triangle can't have two obtuse angles since we can't have two angles more than 90° and the angles of a triangle total only 180°.

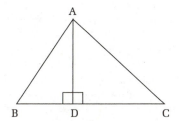

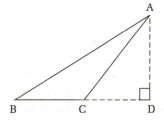

Angle Bisector (again)   BX bisects ∠ABC if ∠ABX = (≅)∠XBC.

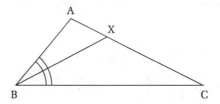

Median: AY   A line segment that is drawn from a vertex to the middle of the opposite side. In other words, AY is a median if BY = (≅)YC.

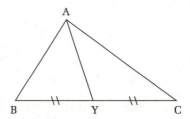

Exterior Angle   ∠AXY is an exterior angle. It is formed by extending one side of a triangle (or any other polygon). The angle on the outside of the triangle (or any other polygon) is that exterior angle. ∠Y and ∠Z are called *remote interior angles* to the exterior angle ∠AXY.

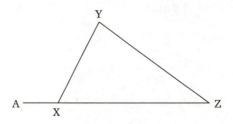

Next we will prove three theorems. The first two are vital not only for this course, but also for the SAT.

Theorem 1   The sum of (the measures) of the angles
of a triangle is 180°.

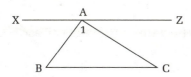

Prove: $(m)\angle B + (m)\angle 1 + (m)\angle C = 180°$

Statement	Reason
1. Draw line XZ through A ∥ to BC.	1. Through a point not on a line, one and only one line can be drawn ∥ to a line.
2. $(m)\angle XAB + (m)\angle 1 + (m)\angle ZAC = 180°$	2. The sum of (the measures of) angles on one side of a line = 180°.
3. But $(m)\angle XAB = (m)\angle B$ aaand $(m)\angle ZAC = (m)\angle C$	3. If two lines are ∥, (the measure of) alternate interior angles are =.
4. $(m)\angle B + (m)\angle 1 + (m)\angle C = 180°$	4. Substitution of 3 into 2.

Theorem 2   The (measure of the) exterior angle
equals the sum of (the measure of) the remote interior
angles.

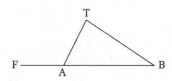

Prove: $(m)\angle FAT = (m)\angle T + (m)\angle B$

Statement	Reason
1. (m)∠FAT + (m)∠TAB = 180°	1. The sum of (the measures of) the angles on one side of a straight line is 180°.
2. (m)∠TAB + (m)∠T + (m)∠B = 180°	2. Previous theorem. The sum of (the measures of) the angles of a triangle is 180°. (*Notice:* some theorems must be proven in sequence.)
3. (m)∠FAT + (m)∠TAB = (m)∠TAB + (m)∠T + (m)∠B	3. Substitute 2 into 1
4. (m)∠TAB = (m)∠TAB	4. Reflexive or identity.
5. (m)∠FAT = (m)∠T + (m)∠B	5. Subtract 3 − 4.

Corollary   (a theorem, usually simple, which is the result of a more important theorem)   An exterior angle is larger than either of the remote interior angle since it is the sum of (the measures of) those two angles.

Let's do the indirect proof I promised you. Here are the steps:

1. Draw the picture as before.

2. Write the given and prove as before.

3. Write all the possibilities.

4. In paragraph form, prove all but one of the possibilities wrong. The remaining one must be correct.

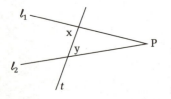

Theorem 3   If two lines form equal (congruent) alternate interior angles, the lines are parallel.

Given transversal t, ℓ_1, ℓ_2 lines: (m)∠x = (m)∠y

Prove: $\ell_1 \parallel \ell_2$

The only possibilities are that either the lines are parallel or the lines are not parallel. Suppose they are not parallel. Then the lines must meet in some point P and we form a triangle. But if this happens $(m)\angle x > (m)\angle y$ since the exterior angle is larger than any remote interior angle. However this is impossible because we assumed that $(m)\angle x = (m)\angle y$. Therefore assuming the lines are not parallel must be wrong. The only other possibility is $l_1 \parallel l_2$. A really good book has a chapter on proofs like this one.

Last, we will do a few numerical examples of theorems 1 and 2. They will probably be on your tests.

EXAMPLE 1—

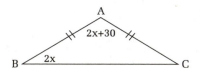

As you see, sides AB and AC are marked the same, so they are equal in length. Therefore the angles opposite are the same. Also the sum of (the measure of) the angles of a triangle ($\triangle$) = 180°. Soooo

$2x + 30 + 2x + 2x = 180 \qquad 6x + 30 = 180$

$6x = 150 \qquad x = 25 \qquad 2x = 50° \qquad$ and $\qquad 2x + 30 = 80°$

The three angles are 50°, 50°, and 80°.

EXAMPLE 2—

In a right triangle the acute angles total 90°. So

$x - 12 + 2x + 3 = 90$ $3x - 9 = 90$ $3x = 99$

$x = 33$ $x - 12 = 21$ and $2x + 3 = 69$

Notice $21 + 69 = 90$.

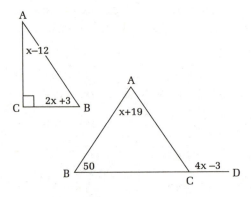

EXAMPLE 3—

The (measure of) the exterior angle equals the sum (of the measures) of the remote interior angles.

$50 + x + 19 = 4x - 3$ $x + 69 = 4x - 3$ $72 = 3x$

$x = 24$ $x + 19 = 43$ $4x - 3 = 93$

Notice $50 + 43 = 93$.

We are now ready to begin the major proof section, the most important chapter or chapters in your book in terms of lasting benefits to you.

CHAPTER 5

SECRETS OF PROVING TRIANGLES CONGRUENT

Proofs like those in this chapter and in at least one later chapter determine how good a book is because these chapters truly teach you to think and write logically. If you spend half the course or more on doing mostly original proofs, then you have truly a value book to be treasured. If you spend only 2 or 3 or 4 weeks on memorized proofs, then either your book is worthless or it really isn't a geometry book (although it may look like a geometry book) or both. We do need one theorem before we get into the chapter.

Theorem Z If two angles of a triangle are = (≅) to two angles of another triangle, the third angles are =(≅).

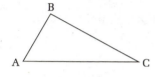
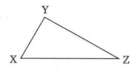

Given: $(m)\angle A = (m)\angle X$

$(m)\angle B = (m)\angle Y$

Prove: $(m)\angle C = (m)\angle Z$

Statement	Reason
1. $(m)\angle A = (m)\angle X$, $(m)\angle B = (m)\angle Y$	1. Given.
2. $(m)\angle A + (m)B = (m)\angle X + (m)\angle Y$	2. Addition.
3. $(m)\angle A + (m)\angle B + (m)\angle C = 180°$	3. Sum of the (measures of the) $\angle$s of a $\triangle = 180°$.
4. $(m)\angle X + (m)\angle Y + (m)\angle Z = 180°$	4. Reason 3.
5. $(m)\angle A + (m)\angle B + (m)\angle C = (m)\angle X + (m)\angle Y + (m)\angle Z$	5. Substitution of 4 into 3.
6. $(m)\angle C = (m)\angle Z$	6. Subtract 5 minus 2.

Not too bad at all.

Now relax and read and do and reread and redo this next set of theorems. You will absolutely positively get the idea. But it may take some time!

There are four major methods of proving triangles congruent. These are theorems which I will not prove. Most books call them postulates so they will not have to prove them or explain why they don't prove them.

Theorem A: Side-Side-Side (abbreviation SSS) If two triangles have three sides, of one equal (congruent) respectively to three sides of another triangle, the triangles are congruent by side-side-side.

On the right are two such triangles. Equal (congruent) side (or angles) are marked the same.

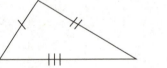 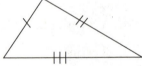

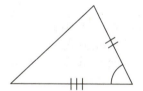 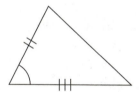

Theorem B: Side-Angle-Side (SAS)   If two triangles have two sides and the included angle of one is equal (congruent) respectively (resp.) to two sides and the included angle of another triangle, then the triangles are congruent by side-angle-side.

   Notice these triangles are backward.

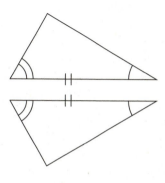

Theorem C: Angle-Side-Angle (ASA)   If two △s have two angles and the included side of one =(≅) resp. to two angles and the included side of another, then the △ s are ≅ by angle-side-angle.

   Notice the triangles can be upside down.

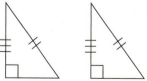

Theorem D: Hypotenuse-Leg (HL)   If a right △ has its hypotenuse and one leg =(≅) resp. to the hypotenuse and leg of another right triangle, then the triangles are congruent by hypotenuse-leg.

   It could have been the other leg also.

NOTE I
There are other ways to show triangles congruent. These are all that are necessary. The others are just repeats, but we will mention them a little later.

NOTE 2
My geometry teacher was the best math teacher I ever had. She forced us to write out, using abbreviations, each reason for each step. I may not do it because it will make the book too long, but you should if you want to be very, very good at these proofs!!!!

NOTE 3

I think one of the worst things books do is spend weeks trying to get you to understand theorems A through D. As we do some of the proofs, you *will* get the idea very shortly.

We are going to do a number of proofs. We will show you how to write them down step by step. *More important,* we will show you how to *think through* the proofs so that you can do original ones easily on your own.

Theorem I   Given isosceles triangle ABC, base AC. BD bisects angle ABC. Prove triangle ABD is congruent to triangle CBD.

Usually the picture is given, especially in the beginning, but we redraw the picture with a ruler, and we have put the given and what is to be proved as before. Here it is.

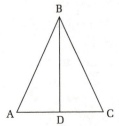

Given: △ABC isosceles, base AC
        BD bisects ∠ABC

Prove: △ABD ≅ △CBD

What You Should See   First, you should see the triangles that are to be proven congruent. If you use a ruler, they will look the same. It really makes the proofs easier if the pictures are good.

  Second, look at the picture. You should be able to tell which parts of the triangles are the same iffff you had already proven the triangles congruent. AB is the same as BC. Angle A and angle C are the same. Angle ABD and angle CBD are the same. AD and DC are the same. BD is in both triangles!!!

Third, in order to prove triangles congruent, we must use theorems A through D.

Last, we can only use the information given, not all the things eyes see are true.

Let's start.

The first step is to write out the first part of the given. The reason is given. Some books have the students write out the entire given in the first step. This is bad, bad, bad!!!!! It hurts understanding and the flow.

The first part looks like this:

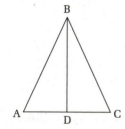

Given: △ABC isosceles, base AC
BD bisects ∠ABC

Prove: △ABD ≅ △CBD

Statement	Reason
1. △ABC isosceles, base AC	1. Given.
A. We have to tell what the uneven side is (the base).	
B. The next step is usually an equal (congruent) statement based on the previous step, and often the reason is explaining the previous step.	
The next line will be	
2. AB = CB (AB ≅ CB)	2. Definition of isosceles triangle [a △ with two sides =(≅)].

Now the proof looks like this:

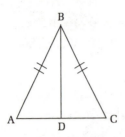

Given: △ABC isosceles, base AC
BD bisects ∠ABC

Prove: △ABD ≅ △CBD

Statement	Reason
1. △ABC isosceles, base AC	1. Given.
2. AB = CB(AB ≅ CB)	2. Definition of isosceles triangle.

We now start another sequence with the second part of the given. At the beginning the sequences will be only one or two steps, but soon there will be more. But they will come naturally.

Again, as before, mark the equal (congruent) parts the same. It looks like this:

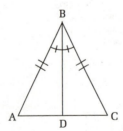

Given: △ABC isosceles, base AC
BD bisects ∠ABC

Prove: △ABD ≅ △CBD

Statement	Reason
1. △ABC isosceles, base AC	1. Given.
2. AB = CB(AB ≅ CB)	2. Definition of isosceles triangle.
3. BD bisects ∠ABC	3. Given.
4. ∠ABD = ∠CBD(∠ABD ≅ ∠CBD)	4. Definition of angle bisector.

What we have so far: We have side-angle. We need
another angle before (for angle-side-angle) or a side after
(for side-angle-side). But we have no more given!!!!!!
What to do, what to do???!! Whenever there is no more
information, we look for vertical angles (the letter X), a
common side between the triangles, or, rarely, a common angle in overlapping triangles. Ahhhh, there is a
common side. The triangles are congruent by side-
angle-side. The finished proof looks like this:

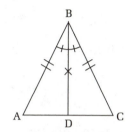

Given: △ABC isosceles, base AC
              BD bisects ∠ABC

Prove: △ABD ≅ △CBD

Statement	Reason
1. ABC isosceles, base AC	1. Given.
S 2. AB = CB(AB ≅ CB)	2. Definition of isosceles triangle.
3. BD bisects ∠ABC	3. Given.
A 4. ∠ABD = ∠CBD(∠ABD ≅ ∠CBD)	4. Definition of angle bisector.
S 5. BD = BD(BD ≅ BD)	5. Identity or reflexive.
6. △ABD ≅ △CBD	6. SAS. If two △ s have two sides and the included ∠ of one =(≅) resp. to two sides and the included ∠ of the other, then the △ s are ≅ by side-angle-side.

1. Read this proof over several times. Actually write
   it out. Try to get the sequence, the flow, the karma.

2. Mark S for side and A for angle to make sure you have everything, until you get good.

3. Read the proof.

4. Some books want the triangles, angles, and sides to be named in exactly the order the objects are the same. I don't care, but some books do. So check this out as you do the proofs.

5. Do number 3 again.

6. Do number 5 again until you understand better.

Let's do another one, but we will not do it in such detail. Instead we will make a third column for comments.

Theorem 2

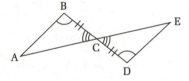

Given: C is the midpoint of BD

AB ∥ ED

Prove: △ABC ≅ △BDC

Statement	Reason	Comments
1. C midpoint of BD	1. Given.	1. Write the first part of the given.
S 2. BC = DC(BC ≅ DC)	2. Definition of midpoint.	2. Explain the given.
3. AB ∥ ED	3. Given.	3. Write the second part of the given. When we have ∥ lines, we look first for the transversal. In this case, we see BD. In this case we see the "Z," for alternate interior angles (the Z is ABCDE). Angles B and D

Statement	Reason	Comments
		are the same. Notice there is another transversal AE, a backward Z, BACED, and another set of alternate interior angles A and E which we are not using in this proof but will in other proofs.
A 4. ∠B = ∠D (∠B ≅ ∠D)	4. If two ∥ lines are cut by a transversal, alternate interior angles are =(≅).	4. We now look for a common side or an X. This time we find the X for vertical angles!!!!
A 5. ∠BCA = ∠DCE (∠BCA ≅ ∠DCE)	5. Vertical angles are =(≅).	5. Nothing to add. Nothing? Impossible!! I talk too much!!!
6. △ABC ≅ △EDC	6. ASA. If two ∠s and the included side = (≅) resp. to two ∠s and the included side of the other, the △ s are ≅ by angle-side-angle.	6. The last statement should always look exacccctly like what you had to prove!!! Keep at it. You'll get it!!!!

Let's try another.

Theorem 3

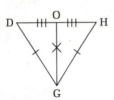

Given: GO bisects DH

GH = DH(GH ≅ DH)

Prove: △DOG ≅ △HOG

Statement	Reason	Comment
1. GO bisects DH	1. Given.	
S 2. DO = OG(DO ≅ OG)	2. Definition of segment bisector.	2. GO cuts DH in half.
S 3. GD = DH(GD ≅ DH)	3. Given.	3. A freebie.
S 4. GO = GO(GO ≅ GO)	4. Identity (reflexive).	4. Again a common side. Go go!!!
5. △DOG ≅ △HOG	5. SSS. If three sides of one are =(≅) resp. to three sides of the other, the △s are ≅ by side-side-side.	5. This is an animal proof.

Getting better? Good. We'll do lots more.

Theorem 4

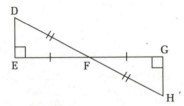

Given: DE ⊥ EG, HG ⊥ EG

F midpoints of both EG and DH

Prove: △DEF ≅ △HGF

Statement	Reason	Comment
1. DE ⊥ EG	1. Given.	1. Even if given is bunched, you separate.

2. DEF is a rt ∠	2. Definition of ⊥.	2. ⊥ is perpendicular (just making sure you know!).
3. HG ⊥ EG	3. Given.	3. We must delay =(≅) cause we only have one angle!
4. HFG is a rt ∠	4. Reason 2.	
rt ∠5. ∠DEF = (≅) ∠HGF)	5. All rt ∠s are =(≅).	5. Whew! Lots of work for one part. Now that we have a right angle, we look for hypotenuse-leg.
6. F midpoint of DH	6. Given.	
hyp. 7. DF =(≅)FH	7. Definition of midpoint.	7. Here are our hypotenuses in both right triangles.
8. F midpoint of EG	8. Given.	8. Another split of the hypothesis.
leg 9. EF =(≅)GF	9. Reason 7.	9. Our legs.
10. △DEF ≅ △HGF	10. If two rt △ s have the hyp. one leg =(≅) resp. to the hyp. and leg of the other, △ s are ≅ by HL, hypotenuse-leg.	10. Hypotenuse-leg is only in a right triangle.

This proof is longer but not harder. Follow the flow, nice and easy.

Lots of notes here:

1. If you are very clever at this point, very, you might see that these triangles could have been proven congruent by SAS using vertical angles and totally avoiding the perpendicular. I did it this way for several reasons: (A) Show HL. (B) Show a slightly longer proof. (C) Point out that proofs with perpendicular tend to be a little longer.

2. While we are on perpendicular, let us see if it is reflexive, symmetric, or transitive. Let a, b, c stand for different lines:

   A. Reflexive a ⊥ a. False. A line can't be perpendicular to itself.

   B. Symmetric: If a ⊥ b, then b ⊥ a. True. See the picture below.

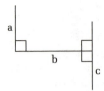

   C. Transitive: If a ⊥ b and b ⊥ c, then a ⊥ c. False. In fact if you are in a plane, a would be *parallel* to c! See the picture above. If you were in space, it might be true. Look at the corner of your classroom. Orrrr. They might be skew lines, not parallel or perpendicular.

   On a true-false test, "sometimes true is called FALSE"!

3. Triangles can be proven congruent by SAA or AAS (side-angle-angle or angle-angle-side). Since two angles are the same, the third angles must also (we proved). So we use ASA. However if your book allows it, use SAA or AAS.

4. Some books have HA, hypotenuse-acute angle. Again, if you have two angles the same (acute angle and right angle), the third angles must be the same, and you can use ASA, which we will do. Again if your book allows it, use HA.

5. But triangles are *not, not, not* congruent by SSA, side-side-angle, the angle not between the two

sides. Let's draw a picture to show two triangles not congruent if there is SSA.

Look at △ ZXW and △ YXW.

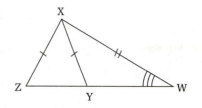

1. ZX is the same as YX.

2. XW is a common side to both.

3. ∠W is common to both.

4. Side-side-angle of one =(≅) SSA, ∠W is not between the corresponding sides, in the others.

5. △ZXW is clearly not the same as △YXW.

Enough of these. Let's try another.

Theorem 5

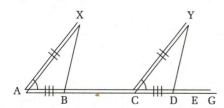

Given: AX = CY(AX ≅ CY)

AX ∥ CY

AC =(≅)BD

Prove: △ABX ≅ △CDY

Thinking

1. We are given a freebie, AX is the same as CY.

2. Parallel lines mean we need a transversal. It is ABCDEG. Here we have corresponding angles. See the "F"? It is crooked on its back.

3. AC and BD are too big and go outside the triangle. But subtracting BC from both, we get AB and CD the same.

4. We have side-angle-side. Ready? Let's write!

Statement	Reason
S1. AX =(≅)CY	1. Given.
2. AX ∥ CY	2. Given.
A3. ∠A =(≅)∠YCD	3. If two ∥ lines are cut by a transversal, corresponding ∠s are =(≅). Notice one angle needs only a single letter; the other angle three letters.
4. AC =(≅)BD	4. Given.
5. BC =(≅)BC	5. Reflexive (identity).
S 6. AB =(≅)CB	6. Subtract 4 − 5.
7. △ABX ≅ △CDY	7. SAS.

Let's try a proof that goes one step beyond congruent triangles. (Sounds like a good name for a sci-fi show.) The step after proving congruent triangles is called . . .

CORRESPONDING PARTS OF CONGRUENT TRIANGLES ARE EQUAL (CONGRUENT)

In order to prove congruent triangles, you need three select parts, each triangle to be the same. After you prove the triangles congruent, all remaining parts are the same. The reason will be corresponding parts. In virtually every proof (in your book, it might be every proof), after proving triangles congruent, the next reason will *always* be corresponding parts of congruent triangles are equal (congruent), abbreviated by CPCTE (CPCTC). I do not like this because many students do not know and understand what this says by using the abbreviation, but I will use it later to save space.

Let's do one!! Yay!!

Theorem 6

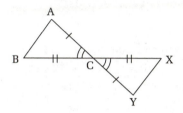

Given: AY and BX bisect each other

Prove: ∠A = (≅)∠Y

Statement	Reason
1. AY bisects BX.	1. Given.
S2. BC =(≅)CX	2. Definition of segment bisector (AY is the cutter).
3. BX bisects AY.	3. Reason 1.
S4. AC =(≅)CY	4. Reason 2 (AY has been cut).
A5. ∠ACB = (≅)∠YCX	5. Vertical angles are equal (congruent).
6. △ ACB ≅ △ YCX	6. SAS. Even though in the proof we wrote SSA, in the picture the angle is between the sides. So we can use SAS.
7. ∠A = (≅)∠Y	7. Corresponding parts of congruent triangles are equal (congruent).

That's not so bad, is it?

   When the proof goes beyond corresponding parts, we have to attack the proof differently. It's the way we *think* about the proof. You think about a proof *backward,* but you write it *forward.*

   Let's try a few.

Theorem 7

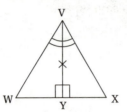

Given: VY ⊥ WX

VY bisects ∠ WVX

Prove: △ WVX is isosceles

How to Think   To show a triangle is isosceles, we must show two sides the same (definition). In this case we must show WV and VX the same. To do that we must have corresponding parts from congruent triangles. To prove congruent triangles, we need to show three parts the same. Let's do it!

Statement	Reason
1. VY ⊥ WX	1. Given.
A 2. ∠VWY = (≅)∠VYX	2. ⊥ s form =(≅)rt ∠ s. Every book is slightly different. You must check your book, especially for this step, to make sure the reasons coincide.
3. VY bisects ∠WVX.	3. Given.
A 4. ∠WVY = (≅)∠XVY	4. Definition of angle bisector.
S 5. VY =(≅)VY	5. Identity (reflexive).
6. △WVY ≅ △XVW	6. ASA. Two notes!!! Notice the side is between the angles; notice we didn't use HL because we didn't have the hypotenuse.
7. WV =(≅)VX	7. Corresponding parts of congruent triangles are equal (congruent). After congruent triangles always comes corresponding parts.
8. △ WVX is isosceles.	8. Definition of isosceles triangle (two sides the same).

That wasn't too bad. Let's try another. Remember, go over each example several times.

Theorem 8

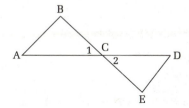

Given: C midpoint of AD and BE

Prove: AB ∥ DE

Thinking   In order to prove ∥ we need to find a Z for alternate interior angles or an F for corresponding angles (others are rare). Here we see the Z (a little rhyme). To show the angles are the same we need corresponding parts after congruent triangles. For congruent triangles, we again need three proper things. In a very short time, this thinking will become natural. Also, in the real world, thinking should be done this way. Let's go!!!

Statement	Reason
1. C is the midpoint of AD.	1. Given.
S 2. AC =(≅) DC	2. Definition of midpoint.
3. C is the midpoint of BE.	3. Given.
S 4. BC =(≅)EC	4. Reason 2.
A 5. ∠1 = (≅)∠2	5. Vertical angles are equal (congruent).
6. △ABC ≅ △DEC	6. SAS.
7. ∠A = (≅)∠D	7. CPCTE(CPCTC).
8. AB ∥ DE	8. If two lines cut by a transversal have alternate interior angles equal (congruent), the lines are parallel.

Not too bad. Let's try a longer type.

Theorem 9: Overlapping Triangles

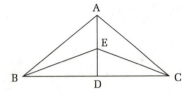

Given: △ ABC isosceles, base BC

AD bisects ∠BAC

Prove: ∠AEB = (≅)∠AEC

Thinking   We want to prove these angles the same. But these angles are in the upper two triangles. Our information is about △ ABD and △ ACD. So in order to show these angles the same by using corresponding parts, we must show the upper triangles congruent. To show the upper ones congruent, we must get corresponding parts from showing the big ones congruent. In each congruence, we must show three things, but some are the same in both. It's not as bad as it sounds, especially after you do seven or eight of them. (I said 7 or 8, not 70 or 80). *Hint:* If you have trouble seeing both sets of triangles, separate them!

Statement	Reason
1. △ ABC isosceles, base BC	1. Given.
S 2. AB = (≅)AC	2. Definition of isosceles triangle.
3. AD bisects ∠BAC.	3. Given.
A 4. ∠BAD = (≅)∠CAD	4. Definition of angle bisector.
S 5. AD = (≅)AD	5. Reflexive (identity).
6. △ADB ≅ △ADC	6. SAS.
7. ∠BAD = (≅)∠CAD	7. CPCTE (CPCTC). Note ∠BAD and ∠BAE are the same angle!!! So are ∠CAD and ∠CAE.
8. AE = (≅)AE	8. Identity (reflexive).

Statement	Reason
9. $\triangle AEB \cong \triangle AEC$	9. SAS from steps 2, 7, and 8.
10. $\angle AEB = (\cong)\angle AEC$	10. Corresponding parts of the upper triangles are equal (congruent).

As with all of these theorems, read them slooooowly, several times. Writing them out helps also. Boy am I tired from writing this chapter. You must be too. That's all for now. We'll revisit later.

CHAPTER 6

SIMILAR FIGURES AND PYTHAGORAS LIVES!!!!

Roughly speaking, similar figures look exactly the same, but are bigger or smaller. I don't think your teacher will let you get away with this explanation. So let's be more formal.

DEFINITION: SIMILAR

Two figures are similar if

   A. Corresponding angles are equal (congruent).

   B. Corresponding sides are in proportion.

Let us give an example.

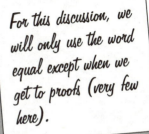

NOTE

For this discussion, we will only use the word equal except when we get to proofs (very few here).

EXAMPLE 1

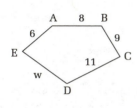

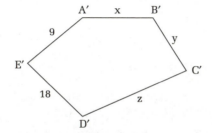

These two figures (pentagons) are similar iffffffffff

1. $\angle A = \angle A'$, $\angle B = \angle B'$, $\angle C = \angle C'$, $\angle D = \angle D'$, $\angle E = \angle E'$ annnnd

2. AB / A'B' = BC / B'C' = CD / C'D' = DE / D'E' = EA / E'A'

In solving for the missing letters w, x, y, and z, we find the corresponding sides in each figure which have numbers. In this case it is AE and A'E'. AE / A'E' = 6 / 9 or 2 / 3 is called the *ratio of similarity*. When we compare any corresponding line segments in this figure, all will be in the ratio of 2/3. Once we say 2/3, it means you always start with the smaller figure.

We say 2/3 = w/18. Cross-multiplying we get 3w = 36. w = 12. Similarly 2/3 = 8/x. x = 12. Also 2/3 = 9/y. So y = 13.5. Lastly 2/3 = 11/z. z = 16.5.

NOTE 1
The notation for similarity is "~." Unfortunately, it is not on my computer; so I have to write them all in by hand.

NOTE 2
Congruent figures are similar, but similar figures don't have to be congruent.

NOTE 3
There are three ways to prove triangles similar:

$$\frac{\approx}{\approx} = \frac{\sim}{\sim}$$

$(m)\angle A = (m)\angle A^1$

A. *Angle-angle similarity:* If two angles of one $\triangle$ are equal (congruent) to two angles of the other $\triangle$, then the triangles are similar by angle-angle similarity.

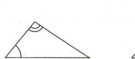 

B. *Side-angle-side similarity:* If two △ s have consecutive sides in proportion and the included
angles are equal (congruent), the triangles are
similar by side-angle-side similarity.

C. *Side-side-side similarity:* If two △ s have their
three sides in the same ratio, then the triangles
are similar by side-side-side similarity.

$$\frac{a}{a'} = \frac{b}{b'} = \frac{c}{c'}$$

 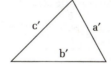

NOTE 4
We will do only a few (informal) proofs using similarity. Few schools do them any more. In any case they
are quite easy.

NOTE 5
In the case of similar figures in general, we need both
angles the same and sides in proportion. In the case of
triangles only, angles are enough (note 3A) or sides are
enough (note 3C).

NOTE 6
We will do almost all similar proofs using angle-angle.

And finally

NOTE 7
Recall that if two angles are the same, so are the third angles. Let's do another example.

EXAMPLE 2

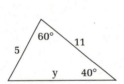 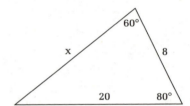

These triangles are similar since in the left triangle the missing angle is 80 and the missing angle in the second triangle is 40, recalling the sum of the (measures of the) angles of a triangle is 180. Also see that these triangles are backward. The ratio of similarity is 5/8.

To get the other sssssidessssss:

$5/8 = y/20$ $8y = 100$ $y = 12.5$ aaand
$5/8 = 11/x$ $5x = 88$ $x = 17.6$

(Remember the side of the smaller triangle goes on top!)

EXAMPLE 3

If BC is ∥ to DE, we have similar △ s, △ABC and △AED, which we will separate for clarity. You may want to do the same. The triangles are similar because of the corresponding angles $\angle 2 = (\cong)\angle 4$ and $\angle 1 = (\cong)\angle 3$. The △s are similar by angle-angle similarity.

8/14 or 4/7 is the ratio of similarity.

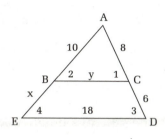

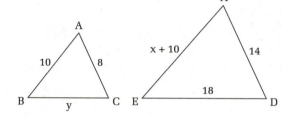

$4/7 = 10/(x + 10)$ $4(x + 10) = 70$ $4x = 30$

$x = 7.5$

(Little right side over big right side = little left over big left.)

$4/7 = y/18$

   (Little right over big right = little base over big base.)

$7y = 72$ $y = 72/7$

EXAMPLE 4—

Backward, upside down. AB ∥ ED   △ ABC~△ DEC since

$\angle A = (\cong)\angle D$ and $\angle B = (\cong)\angle E$

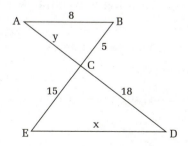

NOTE

In a proportion, you can also do it by saying little right over little base equals big right over big base. We will do this kind of proportion later. (You might try it here to show yourself it works.)

Alternate interior angles, and we again have angle-angle similarity.

AC corresponds to DC!!! See??!! BC corresponds to EC. Of course the bases also correspond.

BC/EC = 5/15 = 1/3, the ratio of similarity.

$1/3 = y/18$ $y = 6$ $1/3 = 8/x$ $x = 24$

This is an important figure that most books spend a few days on, and which leads to one of the most important results in geometry.

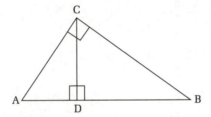

Given right △ ABC with the right angle at C. Drop the altitude to AB at point D. (Recall, altitudes form right angles.) There are three △ s here: small △ DAC, medium △ DCB, and large △ CAB (sort of like papa bear, mama bear, and baby bear triangles).

All three triangles are similar by angle-angle similarity. Why? First, all three triangles have right angles. Angle B is in the small triangle and the big triangle. So they are similar. Angle B is in the medium triangle and the big triangle. So all three triangles are similar!!!

In the small triangle DAC, the short leg is AD, the long leg is CD, the hypotenuse is AC.

In the medium triangle DCB, the short leg is CD, the long leg is BD, the hypotenuse is BC.

In the big triangle CAB, the short leg is AC, the long leg is BC, the hypotenuse is AB.

In chart form, this is. . . .

	Short	Long	Hypotenuse
△ DAC	AD	CD	AC
△ DCB	CD	BD	BC
△ CAB	AC	BC	AB

Study this picture and the chart. Let's do a problem.

EXAMPLE 5—

Solve for x. Look for two △ s where we can set up a proportion with numbers and the letter x only. We see the small and medium triangles fit the description. We say little short over little long equals medium short over medium long

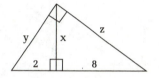

$2/x = x/8$ $x^2 = 16$ $x = 4$

Solve for y. Small and big triangles. Little short over big short equals little hypotenuse over big hypotenuse.

$2/y = y/(2 + 8)$ $y^2 = 20$ $y = \sqrt{20} = 2\sqrt{5}$

Solve for z. We use the medium and large triangles. Medium long over medium hypotenuse equals big long over big hypotenuse.

$8/z = z/(2 + 8)$ $z^2 = 80$ $z = \sqrt{80} = 4\sqrt{5}$

(If you need some help with the radicals, see the Appendix.)

This figure we are studying leads us into one proof of perhaps the most famous theorem in all of mathematics, the *Pythagorean theorem.*

If a and b are legs of a right triangle and the hypotenuse is c, then $c^2 = a^2 + b^2$.

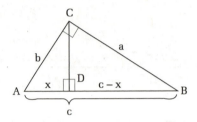

We have already shown these triangles to be similar. We let the letter c be the big hypotenuse AB. If x is one part of the hypotenuse (AD), the other piece, BD, is c − x (the whole minus one piece is the other).

Using the little and big triangles and using small over hypotenuse in each, we get

$x/b = b/c \qquad b^2 = cx$

Using the middle and big triangles and using long over hypotenuse in each, we get

$(c - x)/a = a/c \qquad a^2 = c(c - x)$

Adding, we get

$a^2 + b^2 = cx + c(c - x)$

$a^2 + b^2 = cx + c^2 - cx$

The cx's cancel and we get $a^2 + b^2 = c^2$ or $c^2 = a^2 + b^2$, using the symmetric law. Nice, isn't it? Believe it or not, there are over 100 more proofs of this theorem including three by past presidents of the United States. We actually had smart presidents who liked math.

Let's do a few examples.

EXAMPLE 6—

$a^2 + b^2 = c^2$

$5^2 + 7^2 = x^2$

$x^2 = 74$

$x = \sqrt{74}$

EXAMPLE 7—

$a^2 + b^2 = c^2$

$x^2 + 8^2 = 9^2$ (hypotenuse is always alone)

$x^2 + 64 = 81$

$x^2 = 17$

$x = \sqrt{17}$

EXAMPLE 8—

$$a^2 + b^2 = c^2$$

$$x^2 + (x + 7)^2 = (x + 8)^2$$

$x^2 + x^2 + 14x + 49 = x^2 + 16x + 64$

$2x^2 + 14x + 49 = x^2 + 16x + 64$

$-x^2 - 16x - 64 = -x^2 - 16x - 64$

$x^2 - 2x - 15 = 0$

$(x - 5)(x + 3) = 0$

So x = 5 and x = −3. −3 is a no no since sides can't be negative.

$x = 5$ $x + 7 = 12$ $x + 8 = 13$

The sides are 5, 12, and 13.

This brings us to the Pythagorean triples which you must memorize because they come up over and over and over and you will need them for this course, for the SAT, for trig and beyond, for science, for physics, . . . They are very necessary to speed your work.

Some of the triples are (the hypotenuse is last and always largest):

3, 4, 5 group: 3, 4, 5 ($3^2 + 4^2 = 5^2$); 6, 8, 10; 9, 12, 15; 12, 16, 20; 15, 20, 25.

5, 12, 13 group: 5, 12, 13; 10, 24, 26.

8, 15, 17 group: 8, 15, 17. That's all!

7, 24, 25 group: 7, 24, 25. That's all of them.

There are, of course, many others, actually an infinite number. Two more are 9, 40, 41 and 20, 21, 29. Should you learn these? No! They don't occur often enough. I've learned them because I want to. Math is such fun!!!!!!!!

There are two other right triangles of importance, one comes from a square and the other from an equilateral triangle.

Draw the diagonal to the triangle. What you have is an isosceles right triangle. Since the right angle is 90, the other two are 45 each, the 45°, 45°, 90° isosceles right triangle.

$$s^2 + s^2 = d^2 \qquad d^2 = 2s^2 \cdot \qquad d = s\sqrt{2}$$

In a 45°, 45°, 90° isosceles right triangle:

   A. The legs are the same.

   B. From the leg to hypotenuse, multiply by $\sqrt{2}$.

   C. Hypotenuse to leg? Divide by $\sqrt{2}$. That's it.

EXAMPLE 9—

$BC = 6$

(legs the same)

$AB = 6\sqrt{2}$ (leg times radical 2)

EXAMPLE 10—

$$AC = \frac{10}{\sqrt{2}} \times \frac{\sqrt{2}}{\sqrt{2}} = \frac{10\sqrt{2}}{2} = 5\sqrt{2}$$

BC and AC are the same. That's all there is to it.

Equilateral △ ABC has its altitude (height) drawn. The two smaller triangles are congruent by HL (remember you can't forget the beginning). BD and CD are corresponding parts. Since all sides are "s," BD = 1/2 s.

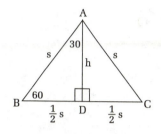

Using old Pythag

$$\left(\frac{1}{2}s\right)^2 + h^2 = s^2 \qquad \frac{1}{4}s^2 + h^2 = 1s^2 \qquad h^2 = \frac{3}{4}s^2$$

$$h = \frac{1}{2}s\sqrt{3}$$

Since all angles of △ ABC are 60° and, by corresponding parts, ∠BAD and ∠CAD are the same, both 30°.

Let us summarize:

Opposite the 90° angle, the hypotenuse is s.

Opposite the 30° angle, the short leg is 1/2 s.

Opposite the 60° angle, the long leg is 1/2s√3.

In rule form:

1. Always get the short leg first if it is not given!!!

2. Short to hypotenuse, multiply by 2.

3. Hypotenuse to short, divide by 2.

4. Short to long, multiply by √3.

5. Long to short, divide by √3.

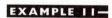**EXAMPLE 11—**

Short (opposite the 30°) is given and is 7.

Hypotenuse is twice the short, 14.

Long is the short x√3 = 7√3.

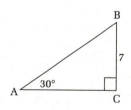

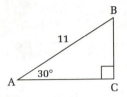

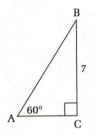

EXAMPLE 12—

Hypotenuse is 11.

Short is hypotenuse divided by 2: 11/2 = 5.5.

Long is $5.5\sqrt{3}$.

EXAMPLE 13—

Long (opposite the 60° angle) is given: 7

Short is long/$\sqrt{3}$ = $7/\sqrt{3} \times \sqrt{3}/\sqrt{3} = 7\sqrt{3}/3$.

Hypotenuse is twice the short = $2 \times 7\sqrt{3}/3 = 14\sqrt{3}/3$.

NOTE 1

When drawing a 30-60-90 triangle, always make the side opposite the 30 really look smaller than opposite the 60. It will really help. Always use a ruler!!!!

NOTE 2

We have shown the height of an equilateral triangle is $s\sqrt{3}/2$.

NOTE 3

The area of an equilateral triangle is $1/2b \times h = 1/2s \times s\sqrt{3}/2 = s^2\sqrt{3}/4$. We will come back to this later.

Enough triangles already. Actually once you get by congruent triangles, the order of a book can be almost anything.

I'll tell you what. Let's do some quadrilaterals now.

CHAPTER 7

QUADRILATERALS SQUARELY DONE

Let's look at *quadrilaterals,* four-sided polygons. When we study them, you must learn their definitions and properties exactly. Teachers love to ask true-false and always-sometimes-never questions about these.

Parallelogram   A quadrilateral with opposite sides parallel. It has the following properties:

1. The opposite sides of a parallelogram are equal (congruent).

2. The opposite angles of a parallelogram are equal (congruent).

3. Consecutive angles of a parallelogram are supplementary.

4. One *diagonal* (a line segment drawn from one vertex to a nonconsecutive vertex) divides the parallelogram into two congruent triangles.

5. The diagonals of a parallelogram bisect each other.

How can you prove a quadrilateral is a parallelogram?

1. By showing opposite sides are parallel.

2. By showing both pairs of opposite sides are equal (congruent).

3. By showing one pair of opposite sides are parallel and equal (congruent).

4. By showing the diagonals bisect each other.

5. By showing the opposite angles are equal (congruent).

Notice: The second five are not all the converses of the first five.

You may have more proofs at this point. Let us do one.

Theorem I

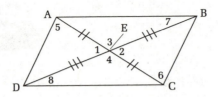

Given: AC bisects BD

BD bisects AC

Prove: ABCD is a parallelogram

Thinking To show it's a parallelogram, we can show the opposite sides are parallel. We have two transversals AC and BD for alternate interior angles, the Zeeee. To get the angles the same, we'll get corresponding parts by proving two pairs of congruent triangles, the upper and lower, and the left and right. OK, let's do it.

Statement	Reason
1. AC bisects BD and BD bisects AC.	1. Given (we're doubling up for brevity).
2. DE = ($\cong$)BE, AE = ($\cong$)CE	2. Definition of segment bisector.
3. $\angle 1$ = ($\cong \angle 2$), $\angle 3$ = ($\cong$)$\angle 4$	3. Vertical angles are equal (congruent).
4. $\triangle$AED $\cong$ $\triangle$CEB, $\triangle$AEB $\cong$ $\triangle$CED	4. SAS in both cases.
5. $\angle 5$ = ($\cong$)$\angle 6$, $\angle 7$ = ($\cong$)$\angle 8$	5. CPCTE (CPCTC).
6. AD$\parallel$CB and AB$\parallel$CD	6. If alternate interior angles are equal (congruent), the lines are parallel.
7. Quadrilateral ABCD is a parallelogram.	7. Definition of a parallelogram (opposite sides of a quadrilateral are parallel).

We will define the remaining quadrilaterals and list the properties, but your book may require you to prove some(not a bad idea since they are similar proofs and not too tough).

NOTE

It's a squashed square.

Rhombus   A parallelogram with all sides equal (congruent).

It has the same properties of a parallelogram + 1.

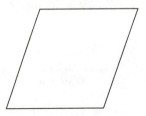

1. Opposite sides are equal (congruent) (actually all sides the same).

2. Opposite angles are equal (congruent).

3. Consecutive angles are supplementary.

4. One diagonal divides it into two congruent triangles.

5. The diagonals bisect each other.

The new kid on the block.

6. The diagonals are also perpendicular.

Rectangle   A parallelogram with right angles.

Properties 1 to 5 of a parallelogram and rhombus.

Not 6.

But 7. The diagonals are equal (congruent).

Square   A rectangle with all equal (congruent) sides. All properties 1 to 7.

Trapezoid   A quadrilateral with exactly one pair of parallel sides.

Isosceles Trapezoid   A trapezoid with nonparallel sides equal (congruent).

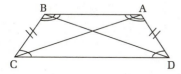

1. The base angles are equal (congruent), top and bottom.

2. The diagonals are equal (congruent).

   AC = (≅) BD

We introduce the following two figures because they will help you on true-false tests, and always-sometimes-never tests.

Kite   Quadrilateral with two pairs of nonopposite sides equal (congruent) and all angles less than 180°. The diagonals are perpendicular to each other.

Arrowhead Quadrilateral with two pairs of nonopposite sides equal (congruent) and one angle more than 180°.

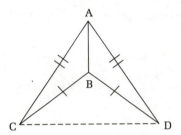

Properties:

1. The diagonals (extended if necessary) are perpendicular.

2. The diagonals *may* be equal (congruent).

NOTE
AB and CD are diagonals.

If you do not see the following picture about quadrilaterals in your book, the book is deficient.

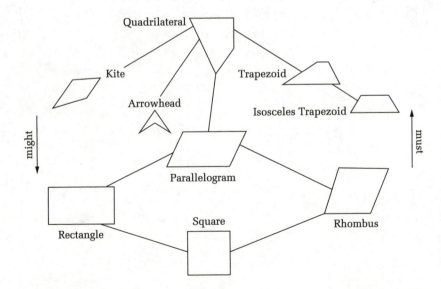

How to Read the Page   If a figure is a parallelogram, it must be a quadrilateral, but might be a rectangle, rhombus, or square.

Must means true. Might means false on a true-false test or sometimes on an always-sometimes-never test.

Sample Question   A quadrilateral with two pairs of equal (congruent) sides is a parallelogram. Answer is sometimes (false on a true-false test). If the opposite sides of a quadrilateral are equal (congruent), it is *always* a parallelogram. This can be touugh. You really must understand and know each definition and each property of each figure. Too many each's. Let's go on.

CHAPTER 8

INTERIOR AND EXTERIOR ANGLES

Another topic, found in most math books and worth talking about, is interior and exterior angle of polygons. But first you should know, if you don't already, the names of polygons.

Once the numbers of sides gets to five or more, we will in general speak of *regular* polygons, polygons with all sides the same and all angles the same.

We of course know the regular three-sided figure: the *equilateral triangle*

And the regular four-sided figure: *square*

Pentagon: five-sided polygon

Hexagon: six-sided polygon

Heptagon: seven-sided polygon

Octagon: eight-sided polygon

Nonagon: guess . . . nine-sided polygon

Decagon: 10-sided polygon

Dodecagon: 12-sided polygon

Pentadecagon: 15-sided polygon

N-gon: n-sided polygon

NOTE

There are n(n − 3) /2 diagonals where n ≥ 3, n is the number of sides.

Now I do the problems on exterior and interior angles a little differently and a little easier than the standard book, so I'll do them both ways and you be the judge. First notice the following:

Three sides: one triangle

Four sides: two triangles with diagonals from one vertex

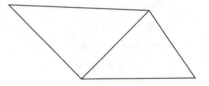

Five sides: three triangles

Six sides: four triangles

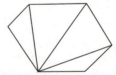

A. In general, the sum of the interior angles of any n-gon is (n − 2)180 degrees since there are (n − 2) triangles and the angles of each triangle add to 180 degrees.

B. An n-gon has n vertices. The sum of the interior plus the exterior angle is 180 degrees times n vertices is n(180) degrees. n(180) − (n − 2)180 = 2(180) = 360 degrees, the sum of the exterior angles.

C. One exterior angle is 360°/n where n is the number of sides.

D. The number of sides n is 360°/1 exterior angle.

E. One interior angle is (n – 2)180/n, but I'll try to show you easier ways.

Remember from the picture that the sum of an interior and an exterior angle is 180.

In doing this kind of problem, *always* try to get the number of sides first. Always use the fact that the sum of the exterior angles is 360°.

EXAMPLE 1—

Tell me about a hexagon:

A. Six sides.

B. 360/6 = 60 degrees, each exterior angle.

C. 180 – 60 = 120 degrees, each interior angle (the supplement of the exterior angle).

D. Total interior: 6(angles) × 120(each interior angle) = 720 degrees   orrrr   4 triangles(6 sides – 2) × 180(degrees in a triangle) = 720 degrees.

Not too bad. Let's try another.

EXAMPLE 2—

Each exterior angle is 10°. Find the rest.

A. Sides = 360/10 = 36 sides.

B. Interior angle is 180 – 10 = 170 degrees.

C. Total interior is (36)(170) or 34(180) = 6120 degrees. Messy but OK. Let's try one more.

EXAMPLE 3—

Total interior is 3240°. Tell me about the figure.

A. method 1: $(n - 2) \, 180 = 3240$

$$180n - 360 = 3240$$

$$180n = 3600$$

$$n = 20 \text{ sides} \qquad \text{orrrrr}$$

A. method 2: $3240/180 = 18$ triangles. 18 (triangles) + 2 = 20 sides.

B. $360/20 = 18$ degrees, one exterior angle. (The 18 is just a coincidence with the other 18.)

C. $180 - 18 = 342$ degrees, each interior angle. OK?? OK!!!

Let's go on to areas and perimeters.

CHAPTER 9

AREAL SEARCH AND SECURING THE PERIMETER

This section is one you should probably be most familiar with, but let's make sure!

Rectangle

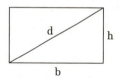

1. Area = base × height $A = b\,h.$

2. Perimeter = 2b + 2h.

3. Diagonal d is found using old Pythagoras
   $d^2 = b^2 + h^2.$

EXAMPLE 1

If the diagonal is 13 feet and the height is 5 feet, find the area and perimeter.

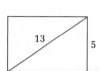

We must get the base first. Sooooo

$b^2 + h^2 = d^2$ $b^2 + 5^2 = 13^2$ $b^2 = 144$
$b = \sqrt{144} = 12$

But of course you knew this was a 5, 12, 13 right triangle.

$A = b\,h = 12 \times 5 = 60$ square feet (ft²)

$p = 2b + 2h = 2(12) + 2(5) = 34$ feet

EXAMPLE 2—

Find the area and perimeter of the following figure. All of the dimensions are in meters.

One way to get the answer is to extend BC until it hits FE at G, forming two rectangles.

$AB = 14$ and $AB = FG$ $FG + GE = FE = 30$
$GE = CD = 16$ $DE = 8$ $DE = CG$
$CG + CB = AF = 20$ So $BC = 12$

Area of left rectangle = b × h = FG × AF = 14 × 20 = 280 square meters.

Area of right rectangle = b × h = GE × DE = 16 × 8 = 128 square meters.

Total area = 280 + 128 = 408 square meters (m²)

Perimeter (add up all outside lengths) = FE + ED + CD + BC + AB + AF = 30 + 8 + 16 + 12 + 14 + 20 = 100 meters.

Square

Rectangle where both base and height are the same.

$A = b \times h = s \times s = s^2$

$p = 2b + 2h = 2s + 2s = 4s$

$d = s\sqrt{2}$ as we showed in the 45-45-90 discussion.

EXAMPLE 3 —

Area of a square is 64 square inches. Find its perimeter and diagonal.

$A = s^2 = 64$ $s = \sqrt{64} = 8$ inches $d = s\sqrt{2} = 8\sqrt{2}$ inches $p = 4s = 4 \times 8 = 32$ inches

NOTE

Squaring comes from a square!!!! Guess where cubing comes from?

Triangle

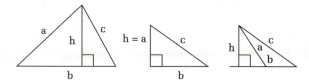

$$A = \frac{1}{2} bh \qquad p = a + b + c$$

h = height, is drawn perpendicular to one of the sides(base) and extended if necessary.

You might think a triangle is half a rectangle. You would be right!!!! Let's show it!

Both regions I and II are the same since the diagonal divides the rectangle (a parallelogram) in half. So a triangle really is half the rectangle. That's why the formula really is 1/2 bh for the area!!!

EXAMPLE 4—

In this right triangle, find a, the area, the perimeter, and y. All measurements are in centimeters.

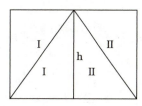

1. $a^2 + 8^2 = 17^2 \qquad a^2 = 225 \qquad a = \sqrt{225} = 15$. Of course you knew this was an 8, 15, 17 right triangle.

2. A = 1/2 b × h = 1/2 8 × 15 = 60 square centimeters (cm²).

3. p = 8 + 15 + 17 = 40 centimeters.

4. There are two ways to get the area: 1/2 8 × 15 = 1/2 17 × y, since y could be considered the height to base 17.

Sooo

120 = 17y   y = 120/17 centimeters

Trapezoid: Area $= 1/2\ h(b_1 + b_2)$

Perimeter $= l_1 + b_1 + l_2 + b_2$

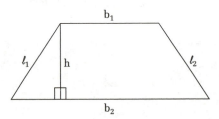

The perimeter is easy; add all the outside lengths. The area is the sum of the area of two triangles. As in the second trapezoid picture, the total area $= 1/2\ b_1 h + 1/2\ b_2 h$. Factoring out one-half h, we get $A = 1/2\ h(b_1 + b_2)$.

EXAMPLE 5—

Find the area and perimeter of the trapezoid if $AB = 12$, $AE = 8$, $BC = 10$ miles.

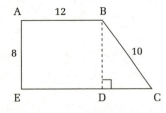

Draw BD perpendicular to CE making ABDE a rectangle. Triangle BDC is a right triangle.

$BC = 10$ and $BD\ (= AE) = 8$

Soooo

$CD = 6$ (6-8-10 right triangle!)

$CE = 12 + 6 = 18$

$l_1 = h = 8$ $b_1 = 12$ $l_2 = 10$ and $b_2 = 18$

$p = 8 + 12 + 10 + 18 = 48$ miles

$A = \dfrac{1}{2}\ 8(12 + 18) = 120$ square miles

Circle

Up to this point we have not dealt with the circle at all. We will deal with it now and several times in the rest of the book, in great detail. For now, let

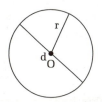

O: the center of the circle.

c = *circumference:* perimeter of the circle.

r = *radius* (plural radii): distance from the center to any point on c.

d = *diameter:* distance from any point on c through O to c on the other side.

r = 1/2 d; d = 2r; c = $2\pi r$ or πd where π is approximately 3.14.

Pi is a number that originated with the Greeks. It is actually an irrational number, a nonrepeating and nonterminating decimal. Recently the most uninteresting book was published: 806 pages of pi written to 2 million places, 3.14159 . . . plus 1,999,995 more places. For the book even to be written, you get an idea of how important pi is in mathematics, verrry important!!!!!

Unfortunately, we cannot derive the formula for the area of the circle because it requires calculus. If you are interested, it can be found in *Calc II for the Clueless.*

EXAMPLE 6—

The diameter of a circle is 8 yards. Find its area and circumference.

$d = 8$ sooo $r = (1/2)8 = 4$

$A = \pi r^2 = \pi\, 4^2 = 16\pi$

I allow my students to leave it in this form, but many want you to do

$A = (3.14)(16) = 50.24 \text{ yd}^2$

$c = 2\pi r = 2(3.14)(4) = 25.12 \text{ yd}$

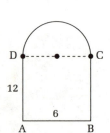

EXAMPLE 7—

We have a Norman window, with a semicircle surmounted on a rectangle. (This is the only time I ever see the word "surmounted.")

If AB = 6 feet and AD = 12 feet, find the area and perimeter.

First draw CD. It is not part of the perimeter, but it helps you see the figure is a rectangle with a semicircle on top.

AB = CD = 12 is the diameter of the semicircle. The semicircle r = 3.

$$A = \frac{1}{2}\pi r^2 + bh = \frac{1}{2}(3.14)(3^2) + (6)(12) = 86.13 \text{ square feet}$$

$$p = BC + AB + AD + \text{a semicircle}$$

$$= 12 + 6 + 12 + (3.14)(3) = 39.42 \text{ feet}$$

Parallelogram

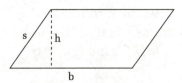

A = bh (the same as a rectangle). p = 2b + 2s.

Rhombus

A = bh (since it is a parallelogram). Also A = 1/2 $d_1 d_2$, half the product of the diagonal. p = 4s, since all sides are the same. Oh, let's do a problem.

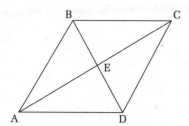

EXAMPLE 8—

Rhombus has sides 13 kilometers and the longer diagonal 24 kilometers, a big rhombus. Find the perimeter and area.

$p = 4s = 52$ kilometers

In the case of a rhombus, the diagonals not only bisect each other but are also perpendicular, forming right triangles. AC = 24. AE = 12, half of AC. AD = 13. Sooo DE = 5, a 5-12-13 right triangle. BD = the other diagonal = 10.

$$A = \frac{1}{2}\, d_1 d_2 = \frac{1}{2}\, (24)(10) = 120 \text{ km}^2$$

Nice problem.

NOTE

If the *ratio of similarity* of two figures is l_1/l_2, the *ratio of their areas* is $(l_1/l_2)^2$.

EXAMPLE 9—

If the ratio of similarity of two triangles is 3/5, and the smaller area is 18 square units, find the area of the larger one.

The proportion is

$$\frac{18}{x} = \left(\frac{3}{5}\right)^2$$

$$\frac{18}{x} = \frac{9}{25}$$

The area of the larger triangle x = 50 square units.

Further note: If we had volumes, the *ratio of the volumes* would be $(l_1/l_2)^3$.

Sector (part of a circle)

AOB is a sector. Its area would be angle/360° πr^2 and the *arc length* s = angle/360° $2\pi r$, since it is that fractional part of the circumference.

EXAMPLE 10—

Let $\angle AOB = 40°$ and OB = 18 inches. For the sector AOB, find its area, arc length, and perimeter.

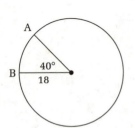

$$A = \frac{40}{360} \pi (18)^2 = 36\pi \text{ square inches}$$

$$s = \frac{40}{360} 2\pi (18) = 4\pi \text{ inches}$$

$$p = s + 2r = 4\pi + 2(18) = 4\pi + 36 \text{ inches}$$

Equilateral Triangle

From before

$$A = \frac{s^2\sqrt{3}}{4} \quad \text{and} \quad p = 3s$$

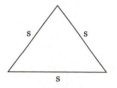

EXAMPLE 11—

Suppose A = $16\sqrt{3}$, find p.

$$16\sqrt{3} = \frac{s^2\sqrt{3}}{4} \quad \text{sooo} \quad s^2\sqrt{3} = 64\sqrt{3} \quad s^2 = 64$$

$$s = 8 \quad p = 3(8) = 24$$

Finally let's talk about a *regular figure.*

The perimeter: the number of sides times s.

Area: Divide the figure into triangles.

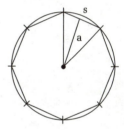

The height of one of the triangles is called an *apothem.*

A = (1/2)(a)(s) times the number of sides. Butttt, s times the number of sides is the perimeter.

So

$$A = (1/2)\ ap$$

Enough of areas and perimeters. Let's do volumes and their surface areas.

CHAPTER 10

VOLUMES AND SURFACE AREAS

This is a chapter that is largely ignored. A week or so should be spent on this chapter because the formulas are needed in later math courses and on the SAT (some of them).

Cube

The SAT is asking more about the shape. So we'll do the cube in great detail. (For more see *SAT Math for the Clueless*.)

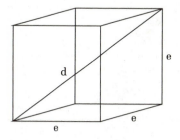

A cube has 12 edges (e), 6 faces, and 8 vertices.

The volume $V = e^3$, read, "e cubed." (Cubing comes from a cube!!!)

There are six faces (surfaces). Each is a square.

The surface area $S = 6e^2$.

The diagonal $d = e\sqrt{3}$.

EXAMPLE 1—

A cube has a volume of 64 cubic feet. Find its surface area and diagonal.

$$e^3 = 64 \qquad e = \sqrt[3]{64} = 4$$

$$S = 6e^2 = 6(4)^2 = 96 \text{ ft}^2 \qquad d = e\sqrt{3} = 4\sqrt{3} \text{ feet}$$

Box

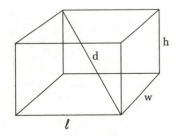

(The real name is rectangular parallelepiped, but that is just toooooo long!)

l: length w: width h: height

Volume $V = l \times w \times h$

Surface Area $S = 2lw + 2\,wh + 2lh$
      (top and   (sides)   (front and
      bottom)   back)

Diagonal $d = \sqrt{l^2 + w^2 + h^2}$, the 3-D Pythagorean theorem

Notice, the cube is a special case of a box.

EXAMPLE 2—

If the length of the box is 12, the width is 3, and the height is 4, find the volume, surface area, and diagonal (everything in meters).

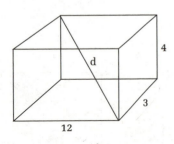

$V = (12)(3)(4) = 144$ cubic meters

$S = 2(12)(3) + 2(3)(4) + 2(12)(4) = 192$ square meters

$d = \sqrt{12^2 + 3^2 + 4^2} = 13$ meters

Cylinder

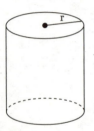

Volume $V = \pi r^2 h$

Surface area $S = 2\pi r^2 + 2\pi rh$

The cylinder has three surfaces:

The top and bottom: $(2\pi r^2)$.

The curved surface: Take a can of soup. Take a knife and caaarefully cut the label straight up or down. You get a rectangle.

The height is the h of the can. The width is the circumference of the can 2π r. Area of rectangle = bh = $2\pi rh$.

I once had a neighbor who wanted to know this formula. All he wanted to know was the answer, but as a teacher I had to tell him how to get it. That is why he moved. Just kidding.

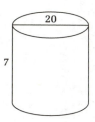

NOTE

When the top and bottom of a figure are the same, the volume will be the area of the base times the height. In this case the base is a circle.

EXAMPLE 3

Find the volume and surface area of a cone whose diameter of the base is 20 feet and whose height is 7 feet. d = 20, r = 10.

$V = \pi (10)(10)7 = 700 \pi$ cubic feet

$S = 2\pi (10)(10) + 2\pi (10)(7) = 340\pi$ square feet

Sphere

$$V = \frac{4}{3} \pi r^3 \qquad S = 4\pi r^2$$

We need *Calc 1 for the Clueless* or *Calc 2 for the Clueless* for motivation and proof.

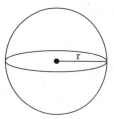

EXAMPLE 4

Find the surface area if V = 36π cubic inches.

$$V = 36\pi = \frac{4}{3} \pi r^3$$

Canceling the pi on each side, and multiplying each side by 3/4, we get $r^3 = 27$. So

$r = 3$ and $S = 4\pi r^2 = 4\pi(3)^2 = 36\pi$ square inches

Notice the two numbers are the same. It is just a coincidence. One is a volume; the other is an area.

Cone

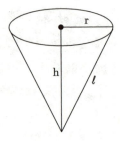

$$V = \frac{1}{3}\pi r^2 h \qquad S = \pi r^2 + \pi r \ell$$

ℓ is called the slant height.

NOTE
If the figure comes to a point, the volume is 1/3 area of the base times the height. Again the base is a circle.

EXAMPLE 5—

Find the volume and surface area of a cone if the radius of the base is 4 and the height is 6. All dimensions are in angstrom units.

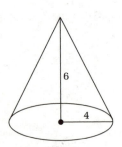

$$V = \frac{1}{3}\pi r^2 h = \frac{1}{3}\pi(4)^2(6) = 32\pi \text{ cubic angstrom units}$$

$$l = \sqrt{4^2 + 6^2} = \sqrt{52} = 2\sqrt{13}$$

$S = \pi r^2 + \pi r \ell = \pi(4^2) + \pi(4)(2\sqrt{13}) = 8\pi(2 + \sqrt{13})$ square angstrom units (the last step by adding and factoring)

Let's do two more figures.

EXAMPLE 6—

We have a prism (or a trough) whose ends are equilateral triangles. If the edge of the equilateral triangle is 10 feet and the length of the prism is 20 feet, find the volume and total surface area.

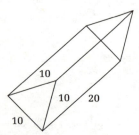

We can consider the equilateral triangle the base and the length the height of the figure.

Volume = area of the base times the height.

$$V = \frac{s^2\sqrt{3}}{4} \times h = \frac{10^2\sqrt{3}}{4}(20) = 500\sqrt{3} \text{ cubic feet}$$

There are five surfaces:

Two equilateral triangles: $S = 2 \times 10^2\sqrt{3}/4 = 50\sqrt{3}$.

Three rectangles = 3 b $\times$ h = $3(10)(20) = 600$ square feet. Total area is $600 + 50\sqrt{3} = 50(12 + \sqrt{3})$ square feet.

EXAMPLE 7—

Given a right pyramid, square base, AB = 6, height of pyramid OP = 4. Find the volume V, slant height PH, edge PC, total surface area S. (A very popular problem.)

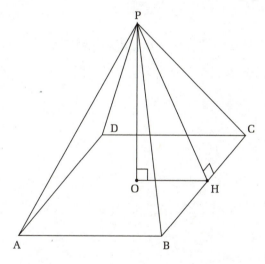

A right pyramid means the height of the pyramid is in the center of the base and of course perpendicular to the base ABCD.

Any figure coming to a point has its volume multiplied by 1/3. Since it is a square base the formula for the volume is

$$V = \frac{1}{3}s^2h = \frac{1}{3}6^2 \times 4 = 48 \text{ cubic units}$$

Since AB = 6, OH which is half of AB = 3.

$PH = \sqrt{OH^2 + OP^2} = \sqrt{3^2 + 4^2} = 5$

A 3, 4, 5 Pythag triple again. If you look closely $\triangle CHP$ is a right triangle, where the angle at H is a right angle. CH = half of BC = half of 6 = 3.

$CP = \sqrt{CH^2 + HP^2}$

$CP = \sqrt{3^2 + 5^2} = \sqrt{34}$

Total surface area is one base plus four triangular sides.

$S = s^2 + 4\frac{1}{2}bl = AB^2 + 2(BC)(PH) = 6^2 + 2(6)(5)$

$= 96$ square units

Now let's go back and look at the circle in greater detail.

First let's take a look at the most common parts of a circle, reviewing from before and naming a few new parts.

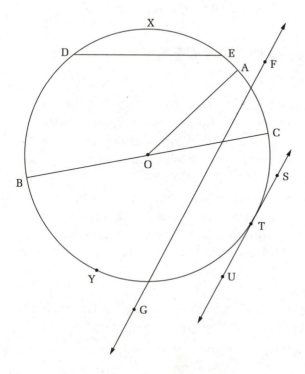

O: center of the circle.

OA: *radius,* as before.

BC: *diameter,* as before.

DE: *chord,* a line segment from one side of the circumference to the other. The diameter is the largest chord.

$\overgroup{DXE}$: *minor arc,* less than half a circle (*arc:* part of the circumference).

$\overgroup{DYE}$: *major arc,* more than half a circle.

$\overgroup{BDC}$: *semicircle,* half a circle.

$\overleftrightarrow{SU}$: *tangent,* a line hitting the circle at one and only one point.

T: *point of tangency.*

$\overleftrightarrow{FG}$: *secant,* a line hitting a circle in two places.

ARCS AND ANGLES WITH ARCS

Arcs can be measured in two ways. The first way, in terms of inches, meters, etc., we have already done. Arcs can also be measured in terms of degrees. As we have said, once around a circle is 360°. An arc that is 1/3 a circle is 1/3 of 360° = 120°.

Books today do almost no proofs in the area. So we will define a few more words, list the theorems, do some problems, and to make me feel a little better, I'll prove a theorem here and there.

DEFINITION I
Central angle, ∠AOC, an angle made with the vertex at the center, O, and whose sides are radii.

DEFINITION 2

Intercepted arc, $\overgroup{AC}$, the arc that $\angle AOC$ cuts off on the circumference.

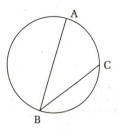

DEFINITION 3

The (measure of the) central angle is equal in degrees, $\overset{\circ}{=}$, to its intercepted arc. And just so I don't forget an important postulate . . .

Postulate I   All radii of the same or equal circles are equal (congruent).

DEFINITION 4

$\angle ABC$: an inscribed angle, an angle formed by two chords with its vertex on the circumference.

Theorem I   The (measure of the) inscribed angle $\angle ABC \overset{\circ}{=} 1/2(m)\overgroup{AC}$. Corollary: All inscribed angles in a semicircle are right angles!!!

Theorem 2   An angle formed by two chords that cross inside the circle is measured by half the sum (of the measures) of their intercepted arcs. Let's prove this one.

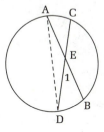

Given: $\angle 1$ formed by intersecting chords AB and CD at E, within the circle

Prove: $(m)\angle 1 \overset{\circ}{=} 1/2\ [(m)\overgroup{BD} + (m)\overgroup{AC}]$

Thought   Draw AD. In $\triangle DEA$, $\angle 1$, an exterior angle is the sum of remote interior angles A and D. Each of the angles is half its intercepted arc. Now let's write it out.

Statement	Reason
1. Chords AB, CD intersect at E	1. Given.
2. Draw AD.	2. One and only one line segment can be drawn between two points.
3. ∠1 exterior angle of △ADE	3. An exterior ∠ of a D is formed by extending one side of a triangle through a vertex.
4. (m)∠1 = (m)∠A + (m)∠D	4. The (measure of the) exterior angle is the sum of (the measure of) the remote interior angles.
5. (m)∠A ≗ 1/2(m)$\widehat{BD}$	5. The (measure of the) inscribed angle ≗ to (the measure of) one-half its intercepted arc.
6. (m)∠1 ≗ 1/2(m)$\widehat{BD}$ + 1/2(m)$\widehat{AC}$	6. Substitution of 5 into 4.
7. (m)∠1 ≗ 1/2((m)$\widehat{BD}$ + (m)$\widehat{AC}$)	7. Factoring.

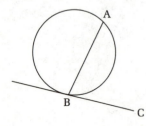

Theorem 3   The (measure of an) angle formed by a tangent and a chord to the point of tangency is equal in degrees to half its intercepted arc.

$$(m)\angle ABC \cong \frac{1}{2}(m)\widehat{AB}$$

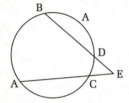

Theorem 4   The angle formed by two secants intersecting outside the circle. The outside angle measures one-half the differences in the arc.

$$(m)\angle E \cong \frac{1}{2}[(m)\widehat{AB} - (m)\widehat{CD}]$$

Theorem 5 The angle formed by a secant and a tangent meeting outside the circle also measures one-half the difference of their intercepted arcs.

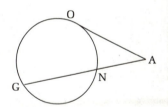

$$(m)\angle A \overset{\circ}{=} \frac{1}{2} \ [(m)\overset{\frown}{GO} - (m)\overset{\frown}{NO}]$$

Theorem 6 If two tangents meet outside the circle, that angle again measures half the difference of the intercepted arcs.

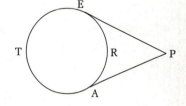

$$(m)\angle P \overset{\circ}{=} \frac{1}{2} \ [(m)\overset{\frown}{ATE} - (m)\overset{\frown}{ARE}]$$

In this case, since both arcs together make a circle, if we let x = the degrees of one of the arcs, the other would be 360 − x.

Let's try some problems. Most of the basic ones are verrry easy.

EXAMPLE 1—

(m)∠AOB = 42°. Find (m)∠ AXB.

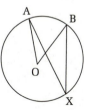

Angle AOB is a central angle and is equal in degrees to its arc AB. Angle AXB is half of its intercepted arc which is also AB. Half of 42 degrees is *21 degrees.*

EXAMPLE 2—

Find (the measure of) x.

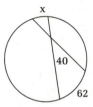

Angles that cross in the circle are half the sum of (the measures of) the arcs. 40 = (1/2)(x + 62)   x + 62 = 80.

x = 18 degrees

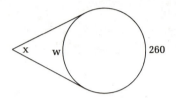

EXAMPLE 3

Find (the measure of) x.

Any angle formed outside the circle measures half the difference in the arcs. But w = 100 = 360 − 260.

$x = (1/2)(260 − 100) = 80$ degrees

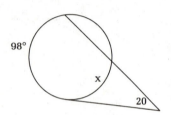

EXAMPLE 4

Find (the measure of) x.

Again, the angle that is outside measures half the difference in the arcs.

$20 = (1/2)(98 − x)$ $40 = 98 − x$ $x = 58$ *degrees*

Most of these basic ones are very easy. However in some books there are very long ones. If you do them systematically they are not too bad. Let's try one. Remember *don't panic!!!!!*

EXAMPLE 5

$(m)\widehat{LM} \stackrel{\circ}{=} 70°$ $(m)\widehat{JK} \stackrel{\circ}{=} 170°$ $(m)\angle UMJ = 40°$

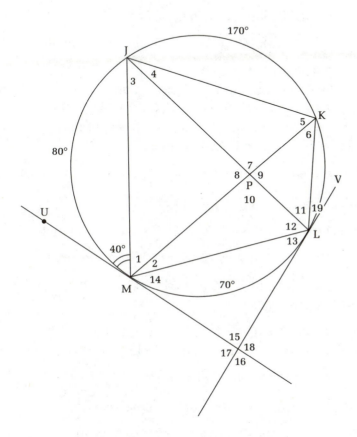

Find all the angles.

Don't panic!!!! It is not as bad as it looks. We first find all the arcs. (m)MJ $\stackrel{\circ}{=}$ 80° since it is double its inscribed angle UMJ (tangent and a secant). The remaining $\widehat{KL}$ = 360 − 70 − 80 − 170 = 40 degrees. Now we can get all remaining angles

$(m)\angle 1 \stackrel{\circ}{=} 1/2(m)\widehat{JK}$ = 1/2 170 = 85°, inscribed angle.

$(m)\angle 2 \stackrel{\circ}{=} 1/2(m)\widehat{KL}$ = 1/2 40 = 20°, inscribed angle.

$(m)\angle 3 \stackrel{\circ}{=} 1/2(m)\overset{\frown}{LM} = 1/2\ 70 = 35°$, inscribed angle.

$(m)\angle 4 \stackrel{\circ}{=} 1/2(m)\overset{\frown}{KL} = 1/2\ 40 = 20°$, inscribed angle.

$(m)\angle 5 \stackrel{\circ}{=} 1/2(m)\overset{\frown}{JM} = 1/2\ 80 = 40°$, inscribed angle.

$(m)\angle 6 \stackrel{\circ}{=} 1/2(m)\overset{\frown}{LM} = 1/2\ 70 = 35°$, inscribed angle.

$(m)\angle 7 \stackrel{\circ}{=} 1/2(m)(\overset{\frown}{JK} + \overset{\frown}{LM}) = 1/2(170 + 70) = 120°$, intersecting chords in the circle   orrr
$(m)\angle 7 + (m)\angle 4 + (m)\angle 5 = (m)\angle 7 + 20 + 40 = 180$. $(m)\angle 7 = 120°$, the sum (of the measures) of the angles of a triangle is 180 degrees.

$(m)\angle 8 = (m)\angle 9 = 60°$, both sups of angle 7.

$(m)\angle 10 = 120°$, vertical of angle 7.

$(m)\angle 11 \stackrel{\circ}{=} 1/2(m)\overset{\frown}{JK} = 1/2\ 170 = 85°$, inscribed angle.

$(m)\angle 12 \stackrel{\circ}{=} 1/2(m)\overset{\frown}{JM} = 1/2\ 80 = 40°$, inscribed angle.

$(m)\angle 13 = (m)\angle 14 \stackrel{\circ}{=} 1/2\ (m)\overset{\frown}{LM} = 1/2\ 70 = 35°$, chord and tangent angle.

$(m)\angle 15 \stackrel{\circ}{=} 1/2(m)(\overset{\frown}{MJKL} - \overset{\frown}{LM}) = 1/2(290 - 70) = 110°$, two tangents meet outside the circle.

$(m)\angle 16 = 110°$, vertical of angle 15.

$(m)\angle 17 = (m)\angle 18 = 70°$, sup of angle 16.

$(m)\angle 19 \stackrel{\circ}{=} 1/2(m)\overset{\frown}{KL} = 1/2\ 40 = 20°$, chord and a tangent.

This really isn't so bad if you analyze slooooowly and carefully and don't panic. Problems like this occur in many geometry books, although most have fewer angles. Redraw this picture and try it yourself!!!!

There are certain lengths that can be proven. Like before, we will give the theorems, given examples, and a proof.

Theorem 7 If two chords cross, the product of each piece is the same.

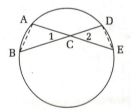

Given: Chords AE and BD meet at C

Prove: AC × CE = BC × CD

Thinking This is a cross-multiplication after finding corresponding parts. Corresponding parts are in proportion after proving similar triangles. The triangles are formed by drawing AB and DE.

Statement	Reason
1. AE, BD chords intersect at C.	1. Given.
2. Draw AB and DE.	2. Through two points one and only one line segment can be drawn.
3. $(m)\angle B \stackrel{\circ}{=} 1/2(m)\overarc{AD}$	3. Inscribed angle = in degrees to half its intercepted arc.
4. $(m)\angle E \stackrel{\circ}{=} 1/2(m)\overarc{AD}$	4. Reason 3.
5. $\angle B = (\cong)\angle E$	5. Substitution (substitution + definition of congruent angles).
6. $\angle 1 = (\cong)\angle 2$	6. Vertical angles are equal (congruent).
7. $\triangle ABC \sim \triangle DEC$	7. Angle-angle similarity.
8. BC/CE = AC/CD	8. Corresponding parts of similar △ s are in proportion.
9. AC × CE = BC × CD	9. Cross-multiplication orrr if equals are multiplied by equals, their products are equal.

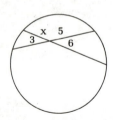

EXAMPLE 6—

Solve for x. Kind of easy.

$6(x) = (3)(5)$ $6x = 15$ $x = 15/6 = 5/2$ or 2.5

EXAMPLE 7—

If chord AB is 10, find each piece. Let x = one piece. Then $10 - x$ is the other piece.

$$x(10 - x) = (5.25)(4)$$

$$10x - x^2 = 21$$

$$x^2 - 10x + 21 = 0$$

$$(x - 7)(x - 3) = 0$$

$$x = 3 \text{ and } 7 \text{ (the two pieces)}$$

Theorem 8   Given AC and AY secants meet at A, then the products of the external piece and the whole secant are equal.

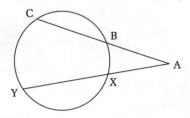

In symbols $AB \times AC = AX \times AY$. The proof is similar to the last proof except for an identity angle instead of verticals. Try to prove it!! I know you can!!

EXAMPLE 8—

Solve for x.

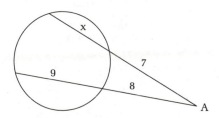

$7(7 + x) = 8(8 + 9)$ $7x + 49 = 136$ $7x = 87$
$x = 87/7 = 12.4$ (approx.)

Theorem 9   Given AD tangent and AC secant meet at A, then the tangent squared is equal to the product of the external piece and the whole secant. In symbols

$AD^2 = AB(AB + BC)$.

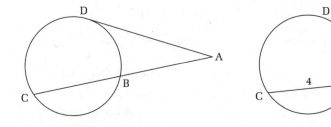

EXAMPLE 9—

Let $AC = x$ $BC = 4$ $AD = \sqrt{32}$. Find x.

$x(x + 4) = (\sqrt{32})^2$ $x^2 + 4x - 32 = 0$
$x^2 + 4x - 32 = 0$ $(x + 8)(x - 4) = 0$

$x = -8$ rejected since on this planet people can't be -8 feet tall.

The answer is $x = 4$.

OTHER THEOREMS

In some books they still do geometric proofs with circles. Hopefully your book still does them, and spends several weeks on them. If it does, the following is very important. In any case you should know the facts.

We will state a couple of postulates, state some theorems, and prove one.

Postulate 1 (again) In a circle or equal circles all radii are equal (congruent).

Postulate 2 (new) Any tangent is perpendicular to the radius drawn to the point of tangency.

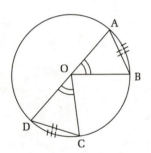

Theorem 9 In a circle or equal circles, if any one of these is true, all are true.

    A. Central angles are equal (congruent).

    B. Their chords are equal (congruent).

    C. Their arcs are equal (congruent).

Outline of Proof Draw in the four radii OA, OB, OC, and OD. Assuming A or B or C is true, we can prove congruent triangles (differently in each case) to show the other two letters are true.

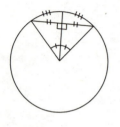

Theorem 10 In a circle or equal circles if any one is true, they all are true.

    A. The central angle is bisected.

    B. Its chord is bisected.

    C. The central angle's intercepted arc is bisected.

    D. The line segment is perpendicular to the chord.

EXAMPLE 10—

Let AB = 12   (m)∠A = 30°   CD ⊥ AB.
Find the radius.

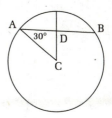

Since CD ⊥ AB, CD also bisects AB.

Since AB = 12, AD = 6. △ DAC is a 30-60-90 right triangle. 6 is leg opposite the 60° angle. To get CD the leg opposite the 30° angle, we divide by square root of 3.

$$CD = \frac{6}{\sqrt{3}} \times \frac{\sqrt{3}}{\sqrt{3}} = \frac{6\sqrt{3}}{3} = 2\sqrt{3}$$

Radius AC = hypotenuse = 2 CD = $4\sqrt{3}$.

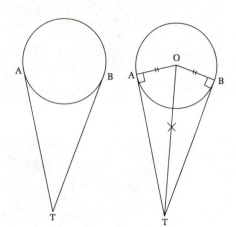

Theorem 11   Tangents from an external point to a circle are equal (congruent).

Given: TA, TB tangents, circle center at O.

Prove: TA = (≅)TB

Statement	Reason
1. TA, TB tangents to circle O.	1. Given.
2. Draw radii OA, OB, and line segment OT.	2. One and only one line segment can be drawn between two points.
3. OB⊥BT, OA⊥AT	3. Postulate: tangents drawn to the radii at point of tangency are perpendicular.
4. ∠OBT = (≅)∠OAT	4. Perpendiculars form equal (congruent) right angles.
5. OA = (≅)OB	5. Postulate: All radii of a circle are equal (congruent).
6. OT = (≅)OT	6. Reflexive (identity)
7. △OBT ≅ △OAT	7. Hypotenuse-leg.
8. TA = (≅)TB	8. Corresponding parts of congruent triangles are equal (congruent).

Notice the proofs are essentially the same, except you use the new properties of circles.

Unless geometry changes back to what it should, in which case there will be a new edition, these are all the congruent triangle proofs for now.

CHAPTER 12

LINES (THE STRAIGHT KIND) AND PARABOLAS I (NOT STRAIGHT)

This chapter is the start of a number of excellent sections (plus one bad one): lines, parabolas, ellipses, hyperbolas, and functions. The only thing is that they shouldn't be in geometry books. They should and must be in algebra books. However, since many geometry books include these topics, we will also.

In elementary algebra, you learned how to graph points and how to graph lines. We will briefly review graphing and then attack the harder problem, finding the equation of a straight line.

STANDARD FORM

When we look at an equation, we should instantly know that it is a straight line. Any equation in the form $Ax + By = C$, where A, B, and C are numbers and A and B are not both equal to 0, is a straight line.

NOTE

In some books, standard form is $Ax + By + C = 0$, so standard form isn't standard. Isn't that funny?!

Let's make sure we understand this by giving some examples that are straight lines and some that are not.

EXAMPLE 1—

A. $3x - 4y = 7$

B. $5x = 9$

C. $x/3 - y/7 = 7$

D. $3/x + 5/y = 9$

E. $xy = 7$

F. $x^2 - 3y = 5$

NOTE

A, B, and C are lines. Coefficients may be negative or fractions or, as in B, $B = 0$. D is not a straight line since the letters are in the bottom. E is not a straight line since the variables are multiplied. F is not a straight line since the exponent of each letter must be 1.

DEFINITION

Slope—The slope (or slant) of a line is defined by $m = (y_2 - y_1)/(x_2 - x_1)$.

NOTES

1. The letter m is always used for the slope.

2. The 1s and 2s are subscripts, standing for point 1 and point 2. The 1 and the 2 mean x_1, x_2, y_1, and y_2 and stand for numbers, not variables, but I will not tell you what they are yet.

3. The calculus notation is $m = \Delta y/\Delta x$, where Δ is the Greek letter delta and means the change in y over the change in x.

4. The y's are always on top.

EXAMPLE 2 ON THE SLOPE—

We will graph the line that joins the points after this example and discuss further. Find the slope between

A. (2,3) and (6,12)

B. (4,–3) and (–1,3)

C. (1,3) and (6,3)

D. (2,5) and (2,8)

SOLUTIONS—

A. (2,3) and (6,12)

$\uparrow\uparrow$ $\uparrow\uparrow$

$x_1 y_1$ $x_2 y_2$

$$m = \frac{y_2 - y_1}{x_2 - x_1} = \frac{12 - 3}{6 - 2} = \frac{9}{4}$$

B. (4,–3) and (–1,3)

$\uparrow\uparrow$ $\uparrow\uparrow$

$x_1 \, y_1$ $x_2 \, y_2$

$$m = \frac{y_2 - y_1}{x_2 - x_1} = \frac{3 - (-3)}{-1 - 4} = \frac{6}{-5}$$

NOTE

In 3 – (–3), the first minus sign is the one in the equation, and the second comes from the fact that y_1 is negative. *Be careful!*

C. (1,3) and (6,3)   m = (3 – 3)/6 – 1) = 0/5 = 0

D. (2,5) and (2,8)   m = (8 – 5)/(2 – 2) = 3/0—
undefined, no slope, or infinite slope

> **NOTE**
>
> *Until you get good at this, label the points just like I did. Also, it does not matter which is point 1 and which is point 2.*

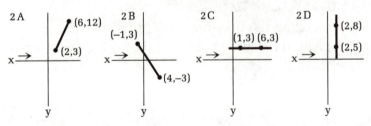

2A. If you walk in the direction of the arrow, when you get to the line and have to walk up, you always have a *positive slope.*

2B. If you have to walk down, you have a *negative slope.*

2C. Horizontal lines have m = 0.

2D. Vertical lines have no slope or infinite slope.

DEFINITION

Intercepts—x intercept, where the line hits the x-axis, is where y = 0. *y intercept* is where x = 0.

EXAMPLE 3—

Given 3x − 4y = 12. Find the intercepts and graph the line.

x-intercept: y = 0. 3x = 12. x = 4. The point is (4,0). (x coordinate is always first.)

y-intercept: x = 0. −4y = 12. y = −3. The point is (0,−3), and the graph is as shown.

NOTE

This is the easiest way to graph a line with two intercepts.

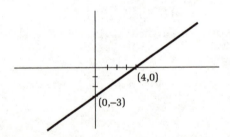

There are three exceptions, the ones with only one intercept:

x = 3 (all vertical lines are x = something).

y = 5 (all horizontal lines are y = something).

y = 3x [where the only intercept is (0,0)]. Pick some other point, say x = 2. So y = 6 and we get the second point (2,6).

Graphs are all included in the picture here.

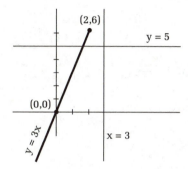

We are almost ready to find the equation of a line. There are four basic forms of a line. The two main ones are the point-slope and the slope-intercept.

DEFINITION

Slope-intercept—y = mx + b. When you solve for y, the coefficient of x is the slope. b is the y intercept—when x = 0, y = b. This is the most common form in high school books.

DEFINITION

Point-slope—m = (y − y$_1$)/(x − x$_1$). You are given the slope and a point (x$_1$,y$_1$). The point-slope is given in different forms in different books. I will try to convince you that this form is the best to use because it eliminates many of the arithmetic fractions, which tend to bother too many of you.

We will now do some problems of this type. Although they are not long problems, most students have difficulty at first. Don't get discouraged if you do.

EXAMPLE 4—

Find the equation of the line through the points (2,3) and (7,11) and write the answer in standard form.

We will do this problem two different ways. I think you may be convinced the point-slope method is better.

Point-slope	**Slope-intercept**
(2,3) and (7,11)	**(2,3) and (7,11)**
$m = \dfrac{(11-3)}{(7-2)} = \dfrac{8}{5}$	$m = \dfrac{(11-3)}{(7-2)} = \dfrac{8}{5}$
$m = \dfrac{y - y_1}{x - x_1}$	$y = mx + b$
$\dfrac{8}{5} = \dfrac{y-3}{x-2}$	$y = \left(\dfrac{8}{5}\right)x + b$
$8(x-2) = 5(y-3)$	$3 = \left(\dfrac{8}{5}\right)(2) + b$
$8x - 16 = 5y - 15$	$3 = \left(\dfrac{16}{5}\right) + b$
$8x - 5y = 1$	$b = 3 - \dfrac{16}{5} = \dfrac{15}{5} - \dfrac{16}{5} = -\dfrac{1}{5}$
	$y = \left(\dfrac{8}{5}\right)x - \dfrac{1}{5}$
	$5y = 8x - 1$
	$-8x + 5y = -1 \text{ or } 8x - 5y = 1$

The reason the point-slope method is easier for most is that the only fractional skill is cross multiplication,* a skill most math students do very well. The slope-intercept method requires a number of fractional skills. About 90% of the time, point-slope is better.

At this stage of your mathematics, all of you should put the lines in standard form to practice your algebra skills. However, to save space and keep the book friendlier, I will leave the answer in either of these two forms.

EXAMPLE 5—

Write the equation of the line with slope 3, y intercept 7.

$y = mx + b$ is the easiest. $y = 3x + 7$.

EXAMPLE 6—

Write the equation of the line with slope –4 and x intercept 9.

Tricky. $m = -4$ and point (9,0). Use point-slope.

$$m = \frac{y - y_1}{x - x_1} \qquad -4 = \frac{y - 0}{x - 9}$$

EXAMPLE 7, PART A—

Find the equation of the line parallel to $3x + 4y = 5$ through (6,7).

EXAMPLE 7, PART B—

Find the equation of the line perpendicular to $3x + 4y = 5$ through (8,–9).

Parallel lines have the same slope. Perpendicular lines have negative reciprocal slopes. In either case we must solve for y.

---

*See the third-to-last line in the point-slope column.

$3x + 4y = 5$. $4y = -3x + 5$. $y = (-3/4)x + 5/4$.

We only care about the coefficient of x, which is the slope. The slope is $-3/4$.

A. Parallel means equal slopes. $m = -3/4$; the point is (6,7). The equation is

$$\frac{-3}{4} = \frac{y - 7}{x - 6}$$

B. Perpendicular means the slope is the negative reciprocal. $m = +4/3$; the point is (8,−9). The equation is

$$\frac{4}{3} = \frac{y - (-9)}{x - 8} \text{ or } \frac{4}{3} = \frac{y + 9}{x - 8}$$

PARABOLA

Later we will do a more complete study of the parabola. We need the basic, standard high school parabola for now. We will study parabolas of the form $y = ax^2 + bx + c$, $a \neq 0$. The coefficient of x^2 determines the parabola's shape.

The low point (or high point), indicated by the letter V, is the vertex, which is why the letter V is usually used.

The x coordinate of the vertex is found by setting x equal to $-b/(2a)$. The y value is gotten by putting the x value into the equation for the parabola. The line through the vertex, the axis of symmetry, is given by $x = -b/(2a)$. The intercepts plus the vertex usually are enough for a fairly good picture. We will do some now.

EXAMPLE 8—

$y = 2x^2 - 7x + 3$

Vertex $x = \dfrac{-b}{(2a)} = \dfrac{-(-7)}{2(2)} = \dfrac{7}{4}$

$y = 2\left(\dfrac{7}{4}\right)^2 - 7\left(\dfrac{7}{4}\right) + 3 = \dfrac{-25}{8}$ $\left(\dfrac{7}{4}, \dfrac{-25}{8}\right)$

Axis of symmetry: $x = \dfrac{-b}{(2a)}$ $x = \dfrac{7}{4}$

y intercept: $x = 0$, $y = 3$ $(0,3)$

x intercepts: $y = 0$

$2x^2 - 7x + 3 = (2x - 1)(x - 3) = 0$

$x = 1/2, 3$ and the intercepts are $(1/2,0)$ and $(3,0)$. The graph opens upward.

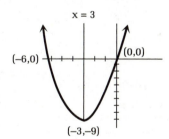

EXAMPLE 9—

$y = x^2 + 6x$

Vertex $x = \dfrac{-b}{(2a)} = \dfrac{-6}{2}(1) = -3$

$y = (-3)^2 + 6(-3) = -9$ $(-3,-9)$

x intercepts: $y = 0$

$x^2 + 6x = x(x + 6) = 0$

$x = 0, -6 (0, 0)$ and $(-6,0)$ are the x intercepts with $(0,0)$ also the y intercept. Picture is up. Axis of symmetry: $x = -3$.

EXAMPLE 10—

$y = 9 - x^2$

Vertex $x = \dfrac{-b}{(2a)} = \dfrac{0}{2}(-1) = 0$

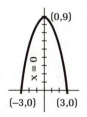

$y = 9$ $(0,-9)$ also y intercept

x intercepts: $y = 0$

$0 = 9 - x^2 = (3 - x)(3 + x)$

$x = 3, -3$; intercepts are $(3,0)$ and $(-3,0)$. Picture is down. Axis of symmetry: $x = 0$.

EXAMPLE 11—

$y = x^2 - 2x + 5$

Vertex $x = -(-2)/2(1) = 1$

$y = 1^2 - 2(1) + 5$ $(1,4)$

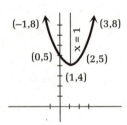

The y intercept is $(0,5)$. $y = x^2 - 2x + 5 = 0$. The quadratic formula gives imaginary roots, so no x intercepts.

To get more points, make a chart. Take two or three x values (integers) just below the vertex and two or three just above.

x	y
−1	8
0	5
1	4
2	5
3	8

Axis of symmetry: $x = 1$. Parabola is up. We are now ready for functions.

CHAPTER 13

DISTANCE FORMULA, MIDPOINT FORMULA, CIRCLE TWO, AND ANALYTIC GEOMETRY PROOFS

DEFINITION

Distance—The distance between two points (x_1, y_1) and (x_2, y_2) is given by the formula
$d = \sqrt{(x_2 - x_1)^2 + (y_2 - y_1)^2}$.

DEFINITION

Midpoint—The average (mean) between two points is $[(x_1 + x_2)/2, (y_1 + y_2)/2]$.

EXAMPLE I

Given $(3, -5)$ and $(9, -1)$. Find the length of the line segment between the points and the midpoint of that line segment.

$$d = \sqrt{\{(9 - 3)^2 + [-1 - (-5)]^2\}} = \sqrt{(36 + 16)} = \sqrt{52}$$

$$= \sqrt{2 \cdot 2 \cdot 13} = 2\sqrt{13}$$

Midpoint is $\{(3 + 9)/2, [-1 + (-5)]/2\} = (6, -3)$.

DEFINITION

Circle—Set of all points (x, y) at a distance r (don't tell anyone, but r stands for radius) from a given point (h, k) called the center of the circle.

The equation of a circle is the square of the distance formula $(x - h)^2 + (y - k)^2 = r^2$.

EXAMPLE 2—

Find the center and radius of $(x - 3)^2 + (y + 5)^2 = 7$.

C(3,–5) which is the opposite sign, and the radius = $\sqrt{7}$.

EXAMPLE 3—

Find r and C for the circle $2x^2 + 2y^2 - 12x + 6y + 8 = 0$.

First, we know it's a circle, since the coefficients of x^2 and y^2 are the same, as long as the radius turns out to be a positive number.

Coefficients of x^2 and y^2 must be 1.	$2x^2 + 2y^2 - 12x + 6y + 8 = 0$ or
Group x's and y's together, constant to other side.	$x^2 + y^2 - 6x + 3y + 4 = 0$ or $x^2 - 6x + y^2 + 3y = -4$ or
Complete the square and add the term(s) to each side.	$x^2 - 6x + (-6/2)^2 + y^2 + 3y + (3/2)^2 = (-6/2)^2 + (3/2)^2 - 4$ or
Factor and do the arithmetic.	$(x - 3)^2 + (y + 3/2)^2 = 29/4$

The center is (3,–3/2) and the radius is $\sqrt{\dfrac{29}{4}} = \dfrac{\sqrt{29}}{2}$.

ANALYTIC GEOMETRY PROOFS

Analytic geometry, or algebraic geometry, used to be part of all geometry books. I feel compelled to show you two examples of what was and hopefully will be again in your books.

The picture we draw will be used for both proofs.

Theorem 1 Prove the diagonals of a parallelogram bisect each other.

We must draw a parallelogram with the x and y axes (plural of axis) the easiest way possible. One vertex is at the origin, (0,0). Another vertex is on the x axis, point (a,0), length of side is "a." A third vertex is (b,c). The 4th vertex: the side is parallel to the one on the x axis and the same length. Height is the same and the x value is "a" units more. The point is (a + b,c). (This part takes practice.)

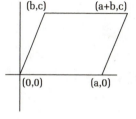

To show the diagonals bisect each other, we can show both diagonals have the same midpoint!!!! The midpoint between (0,0) and (a + b,c) = (a + b/2,c + 0/2). The midpoint between (a,0) and (b,c) = (a + b/2,c + 0/2). Since they are the same, the diagonals bisect each other. The proof itself is easy if (a big if) you set it up and see what has to be done.

Theorem 2 If the diagonals of a parallelogram are equal (congruent), the figure is a rectangle.

Lengths the same in algebra mean distance formula. We will use the *square* of distance, since if we don't, the first step will be to square both sides. Then we will simplify and see what happens.

Diagonals the same means

$$(x_2 - x_1)^2 + (y_2 - y_1)^2 = (x_4 - x_3)^2 + (y_4 - y_3)^2$$

$$[(a + b) - 0]^2 + (c - 0)^2 = (b - a)^2 + (c - 0)^2$$

Multiplying out, we get

$$a^2 + 2ab + b^2 + c^2 = b^2 - 2ab + a^2 + c^2$$

All square terms cancel. Sooo we get

$$2ab = -2ab \text{ or } 4ab = 0.$$

Dividing by 4 we get that ab = 0. This means either a = 0 or b = 0.

Looking at the picture "a" could not be 0. There would be no parallelogram. So the length b = 0. But if b = 0, b must be on the y axis, and the parallelogram is a rectangle!!!

These are neat proofs, not that easy and not like the proofs before. They improve your brain. So I hope they'll return soon!!!

CHAPTER 14

FUNCTIONS, TRANSLATIONS, STRETCHES, CONTRACTIONS, FLIPS

One of the most neglected topics in high school is the study of functions. In this book there are three rather lengthy chapters directly related to functions and several others that are indirectly related. There are two reasons for this: functions are important, and most calculus courses assume you know this topic almost perfectly, an unrealistic assumption. So let's get started at the beginning.

Function: Given a set D. To each element in D, we assign one and only one element.

EXAMPLE 1—

Does the picture here represent a function? The answer is yes. 1 goes into a, 2 goes into 3, 3 goes into 3, and 4 goes into *pig*. Each element in D is assigned one and only one element.

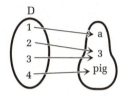

The next example will show what is not a function. But let us talk a little more about this example. The set D is called the *domain.* We usually think about x values when we think about the domain. This is not necessarily true, but it is true in nearly all high school and college courses, so we will assume it.

There is a second set that arises. It is not part of the definition. However, it is always there. It is called the *range*. Notice that the domain and the range can contain the same thing (the number 3) or vastly different things (3 and *pig*). However, in math, we deal mostly with numbers and letters. The rule (the arrows) is called the *map* or *mapping*. 1 is mapped into a; 2 is mapped into 3; 3 is mapped into 3; and 4 is mapped into *pig*.

FUNCTIONAL NOTATION

The rule is usually given in a different form: f(1) = a (read "f of 1 equals a"); f(2) = 3; f(3) = 3; and f(4) = *pig*.

NOTE 1
When we think of the range, we will think of the y values, although again this is not necessarily true.

NOTE 2
We cannot always draw pictures of functions, and we will give more realistic examples after we give an example of something that is not a function.

EXAMPLE 2

The picture here does not represent a function, since 1 is assigned two values, a and d.

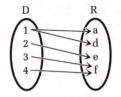

EXAMPLE 3

Let $f(x) = x^2 + 4x + 7$. D = {1, −3, 10}.

$f(1) = (1)^2 + 4(1) + 7 = 12$

$f(−3) = (−3)^2 + 4(−3) + 7 = 4$

$f(10) = (10)^2 + 4(10) + 7 = 147$

The range would be {4,12,147}. If we graphed these points, we would graph (1,12), (−3,4), and (10,147).

NOTE

Instead of graphing points (x,y), we are graphing points (x,f(x)). For our purposes, the notation is different, but the meanings are the same.

EXAMPLE 4—

Let $g(x) = x^2 - 5x - 9$. D = {4,0,−3,a^4,x + h}. Find the elements in the range.

This is a pretty crazy example, but there are reasons to do it.

$$g(4) = (4)^2 - 5(4) - 9 = -13$$

$$g(0) = 0^2 - 5(0) - 9 = -9$$

$$g(-3) = (-3)^2 - 5(-3) - 9 = 15$$

$$g(a^4) = (a^4)^2 - 5a^4 - 9 = a^8 - 5a^4 - 9$$

$$g(x + h) = (x + h)^2 - 5(x + h) - 9$$

$$= x^2 + 2xh + h^2 - 5x - 5h - 9$$

The range is {−13,−9,15,$a^8 - 5a^4 - 9$,$x^2 + 2xh + h^2 - 5x - 5h - 9$}.

TRANSLATIONS, STRETCHES, CONTRACTIONS, FLIPS

If there was ever a more useless topic than this, I want to know it. It occurs in either this course or algebra 2. It not only wastes time, replacing something, anything, more valuable, it also can hurt you because when you sketch sin x or cos x in trig, the translating, stretching, contracting, and flipping is done much differently. But because it is in your book, here goes.

Suppose f(x) looks like this.

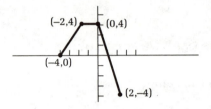

EXAMPLE 1—

f(x) + 3 would be 3 units up in the y direction and would look like this.

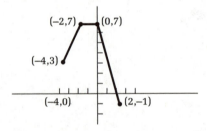

EXAMPLE 2—

f(x) − 4 is 4 units down. (Look at f(x), the original, first.)

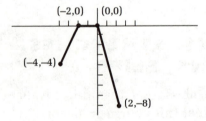

EXAMPLE 3—

2f(x): twice as high (or twice as low). (Points on the x-axis stay the same.)

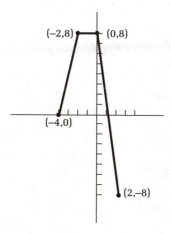

EXAMPLE 4—

(1/4)f(x): 1/4 as high or as low.

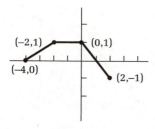

EXAMPLE 5

–f(x): upside down. Flip the x-axis. Top becomes bottom and bottom becomes top.

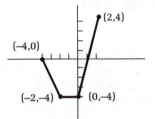

EXAMPLE 6

f(x − 3): three units to the right.

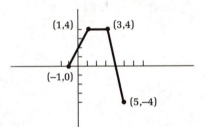

EXAMPLE 7

f(x + 5): five units to the left.

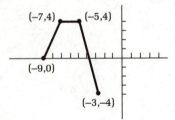

EXAMPLE 8—

f(–x): flip the y-axis. Right becomes left and left
becomes right.

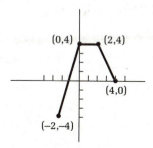

EXAMPLE 9—

f(2x): compact, twice as close. (Points in the y-axis stay
the same.)

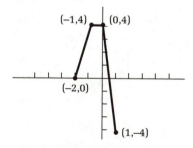

EXAMPLE 10—

$f\left(\dfrac{1}{3}x\right)$: stretch three times.

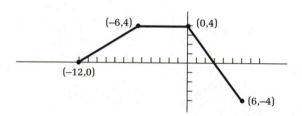

There can be combos: $y = -2f(x - 5) + 3$. Order of operations:

1. Inside parenthesis, five units to the right.

2. Multiplying the function -2: twice as high, upside down.

3. Up or down $+3$ up 3. Let's draw it in stages.

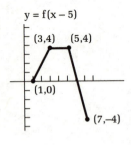

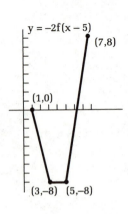

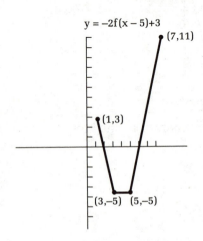

CHAPTER 15

PARABOLAS II, ELLIPSES, AND HYPERBOLAS

We will now discuss these three curves the way they are discussed in some precalc and algebra 2 books.

The important way to study these curves is to relate the equation to the picture. If you do this, this entire chapter will become much easier.

DEFINITION

Parabola—The set of all points equidistant from a point, called a *focus,* and a line, called the *directrix.* F is the focus. The point V, the *vertex,* is the closest point to the directrix.

NOTE

According to the definition $FV = VR$, $FP_1 = P_1R_1$, $FP_2 = P_2R_2$, etc.

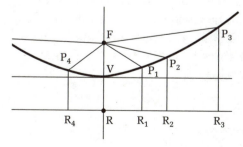

Let's do this development algebraically.

Let the vertex V be at (0,0) and the focus F be at (0,c). The directrix is at $y = -c$. Let P(x,y) be any point on the parabola. The definition of the parabola says FP = PQ. Everything on PQ has the same x value, and everything on $y = -c$ has the same y value. Point Q has to be (x,−c). On PQ, since the x values are the same, the length of the line is $y - (-c)$. Using the distance formula to get PF and setting the two segments squared equal to each other, we get . . . $(x - 0)^2 + (y - c)^2 = (y + c)^2$. Squaring, we get $x^2 + y^2 - 2cy + c^2 = y^2 + 2cy + c^2$. Simplifying, we get $x^2 = 4cy$.

Here is a chart that will be helpful in relating the equation to the picture.

Vertex	Focus	Directrix	Equation	Picture	Comment
(0,0)	(0,c)	$y = -c$	$x^2 = 4cy$		The original derivation
(0,0)	(0,−c)	$y = c$	$x^2 = -4cy$		y replaced by −y
(0,0)	(c,0)	$x = -c$	$y^2 = 4cx$		x,y interchange in top line
(0,0)	(−c,0)	$x = c$	$y^2 = -4cx$		x replaced by −x in the above line

EXAMPLE 1—

$y^2 = -7x$. Sketch; label vertex, focus, directrix.

The chart tells us the picture is the last line. Now let 4c equal 7, ignoring the minus sign. c = 7/4. Vertex is (0,0). Focus is (−7/4,0) because it is on the negative x axis. The directrix is x = +7/4, positive because it is to the right and x = 7/4 since it is a vertical line. Easy, isn't it?

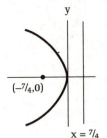

EXAMPLE 2—

Sketch $(y - 3)^2 = -7(x + 2)$.

To understand this curve, we look at the difference between $x^2 + y^2 = 25$ and $(x - 3)^2 + (y + 6)^2 = 25$. Has the shape changed? No. Has the radius changed? No. The only thing that has changed is its position. The center is now at the point $(3,-6)$ instead of at the point $(0,0)$.

In the case of our little parabola, it is the vertex that has changed. $V = (-2,3)$. (Remember x is always first.) 4c is still equal to 7. $c = 7/4$, but F is $(-2 - 7/4, 3)$, 7/4 to the left of the vertex. The directrix is $x = -2 + 7/4$. I do not do the arithmetic here so that you know where the numbers come from.

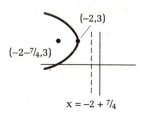

EXAMPLE 3—

Sketch the parabola $2x^2 + 8x + 6y + 10 = 0$.

$2x^2 + 8x + 6y + 10 = 0$

Divide by coefficient of x^2 or y^2.

$x^2 + 4x + 3y + 5 = 0$

On the left get all the terms related to the square term, everything else on the other side.

$x^2 + 4x = -3y - 5$

Complete the square; add to both sides.

$x^2 + 4x + 4 = -3y - 5 + 4$

Factor; do the arithmetic.

$(x + 2)^2 = -3y - 1$

Weird last step. No matter what the coefficient on the right, factor it all out, even if it results in a fraction in the parentheses.

$(x + 2)^2 = -3\left(y + \dfrac{1}{3}\right)$

Sketch; the picture is on the following page. $V(-2,-1/3)$. $4c = 3$. $c = 3/4$. $F(-2,-1/3 - 3/4)$. Directrix $y = -1/3 + 3/4$.

$y = -1/3 + 3/4$

$v(-2, -1/3)$

$F(-2, -1/3 -3/4)$

DEFINITION

Algebraically, the *ellipse* is defined as $PF_1 + PF_2 = 2a$. $2a > 2c$. $2c$ is the distance between the *foci*. a will be determined later. P is any point on the ellipse. To paraphrase the definition, given two points called foci, we draw a line from one of the points to any point on the curve and then from that point on the curve to the other focus; if the sum of the two lengths always adds to the same number, $2a$, the figure formed will be an ellipse.

I know you'd all desperately like to draw an ellipse. Take a nonelastic string. Attach thumbtacks at either end. Take a pencil and stretch the string as far as it goes. Go 360°. You will trace an ellipse.

Some of you have seen the equation of an ellipse, but few of you have seen its derivation. It is an excellent algebraic exercise for you to try. You will see there is a lot of algebra behind a very simple equation.

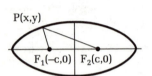

Use the distance formula.

$$PF_1 + PF_2 = 2a$$

Isolate one square root and square both sides.

$$\sqrt{(x - (-c))^2 + (y - 0)^2}$$
$$+ \sqrt{(x - c)^2 + (y - 0)^2} = 2a$$

Do the algebra.

$$[\sqrt{(x + c)^2 + y^2}]^2 = [2a - \sqrt{(x - c)^2 + y^2}]^2$$

Combine like terms; isolate the radical.

$$x^2 + 2cx + c^2 + y^2 = 4a^2 + x^2 - 2cx + c^2 + y^2$$
$$- 4a\sqrt{(x - c)^2 + y^2}$$

Divide by 4; again square both sides.

$$4a\sqrt{(x - c)^2 + y^2} = 4a^2 - 4cx$$

Do the algebra.

$$[a\sqrt{(x - c)^2 + y^2}]^2 = (4a^2 - 4cx)^2$$

$$a^2(x^2 - 2cx + c^2 + y^2) = a^4 - 2a^2cx + c^2x^2$$

$$\text{or} \quad a^2x^2 - c^2x^2 + a^2y^2 = a^4 - a^2c^2$$

$$\frac{(a^2 - c^2)x^2}{(a^2 - c^2)a^2} + \frac{a^2y^2}{a^2(a^2 - c^2)} = \frac{a^2(a^2 - c^2)}{a^2(a^2 - c^2)}$$

Factor out x^2 on the left two terms and a^2 from the right two terms; divide by $(a^2 - c^2) \times a^2$.

We get

$$\frac{x^2}{a^2} + \frac{y^2}{a^2 - c^2} = 1 \qquad \text{or} \qquad \frac{x^2}{a^2} + \frac{y^2}{b^2} = 1$$

letting $a^2 - c^2 = b^2$. Whew!!!!!!

We are still not finished. Let's find out what a is and what b is. $F_1P + PF_2 = 2a$, where P is any point on the ellipse. Let T be the point P. $F_1T + TF_2 = 2a$. By symmetry, $F_1T = TF_2$. So $F_1T = TF_2 = a$. Since $a^2 - c^2 = b^2$, OT = OT' = b. The coordinates of T are (0,b); T' is (0,−b).

We would like to find the coordinates of U (and U'), but the letters a, b, and c are used up. Oh well, let's see what happens. $F_2U + UF_1 = 2a$. $F_2U = x - c$. $UF_1 = x + c$. $x - c + x + c = 2a$. So $2x = 2a$; $x = a$. U is (a,0); U' is (−a,0).

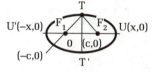

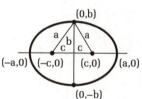

c = half the distance between the foci

b = length of the semiminor axis ("semi" means half; "minor" means smaller; "axis" means line)

a = length of semimajor axis = distance from a focus to a minor vertex

(±a,0)—major vertices

(0,±b)—minor vertices

(±c,0)—foci, always located on the major axis

Although the derivation was long, sketching should be short.

EXAMPLE 4—

Sketch $x^2/7 + y^2/5 = 1$.

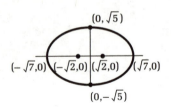

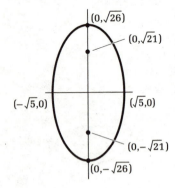

In the case of the ellipse, the longer axis is indicated by the larger number under x^2 or y^2. Try not to remember a or b. Remember the picture! This is longer in the x direction. $y = 0$, major vertices $(\pm\sqrt{7},0)$; $x = 0$, minor vertices $(0,\pm\sqrt{5})$. $c = \sqrt{(7-5)}$. The foci are $(\pm\sqrt{2},0)$.

EXAMPLE 5—

$$\frac{x^2}{5} + \frac{y^2}{26} = 1$$

Longer in the y direction. Major vertices: $(0,\pm 26^{1/2})$. Minor vertices: $(0,\pm\sqrt{5})$. $c = \sqrt{(26-5)}$. Foci: $(0,\pm\sqrt{21})$. Foci are always on the longer axis.

EXAMPLE 6—

$$\frac{(x-6)^2}{7} + \frac{(y+4)^2}{5} = 1$$

This is the same basic example as Example 4, except the middle is no longer at (0,0). It is at the point (6,–4). Major vertices: $(6 \pm \sqrt{7},-4)$. Minor vertices: $(6,-4 \pm \sqrt{5})$. Foci: $(6 \pm \sqrt{2},-4)$.

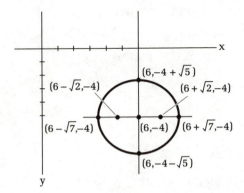

NOTE
The numbers found in Example 4 are added and subtracted from the appropriate coordinate of the center

(–6,–4). Also note that weird numbers were purposely chosen so that you could see where the numbers came from.

EXAMPLE 6 REVISITED—

$$\frac{(x-6)^2}{7} + \frac{(y+4)^2}{5} = 1$$

Just a few months after my last book went to press, I realized a second way to find the vertices that would make it much clearer for you, but my editor wouldn't change the book. So now I finally have a chance to show you.

The center is again at the point (6,–4). The vertices are directly east-west or north-south of (6,–4). All points east-west of (6,–4) must have the same y value, y = –4. Soooooo

$$\frac{(x-6)^2}{7} + \frac{(-4+4)^2}{5} = 1$$

$$\frac{(x-6)^2}{7} + 0 = 1$$

$$(x-6)^2 = 7$$

$$x - 6 = \pm\sqrt{7}$$

Soooooo x = 6 $\pm$ $\sqrt{7}$ and the major vertices are (6 $\pm$ $\sqrt{7}$,–4). Points north-south of (6,–4) have the same x values, x = 6.

$$\frac{(6-6)^2}{7} + \frac{(y+4)^2}{5} = 1$$

$$0 + \frac{(y+4)^2}{5} = 1$$

$$(y+4)^2 = 5$$

$$y + 4 = \pm\sqrt{5}$$

Therefore $y = -4 \pm \sqrt{5}$ and the minor vertices are $(6, -4 \pm \sqrt{5})$. For the same reasons, the foci, always on the major axis, are $(6 \pm \sqrt{2}, -4)$. The sketch is, of course, the same.

EXAMPLE 7—

Sketch and discuss $4x^2 + 5y^2 + 30y - 40x + 45 = 0$.

Like the parabola and circle, we must complete the square, only a little differently.

Group the x and y terms; number to the other side.

$$4x^2 + 5y^2 + 30y - 40x + 45 = 0$$

$$4x^2 - 40x + 5y^2 + 30y = -45$$

Factor out coefficients of x² and y²; complete the square in the parentheses; add the number term inside the parentheses multiplied by the number outside the parentheses to each side—with both x and y. Then do the arithmetic and divide by 100 to get 1 on the right.

$$4\left[x^2 - 10x + \left(\frac{-10}{2}\right)^2\right] + 5\left[y^2 + 6y + \left(\frac{6}{2}\right)^2\right]$$

$$= -45 + 4\left(\frac{-10}{2}\right)^2 + 5\left(\frac{6}{2}\right)^2$$

$$\frac{4(x-5)^2}{100} + \frac{5(y+3)^2}{100} = \frac{100}{100}$$

or $\quad \dfrac{(x-5)^2}{25} + \dfrac{(y+3)^2}{20} = 1$

The center is $(5, -3)$. Under the $(x-5)^2$ term is larger; the ellipse is long in the x direction. $\sqrt{25}$ units are to the left and right of the center. Major vertices are $(5 \pm \sqrt{25}, -3)$.

Under the $(y + 3)^2$, $\sqrt{20}$ above and below the center.
Minor vertices are $(5,-3 \pm \sqrt{20})$. $c = \sqrt{(25 - 20)}$. Foci,
on the larger axis, are $(5 \pm \sqrt{5},-3)$.

Of course you should use 5 instead of $\sqrt{25}$, but I left
$\sqrt{25}$ to show you where the 5 came from.

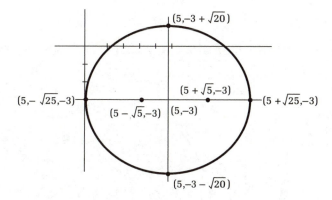

DEFINITION

Hyperbola—set of all points P such that $F_1P - PF_2 = 2a$.
The derivation is exactly the same as for the ellipse.
Once is enough!!!!! The equation is $x^2/a^2 - y^2/b^2 = 1$,
where $a^2 + b^2 = c^2$. $(\pm a,0)$—transverse vertices. $(\pm c,0)$—
foci. Asymptotes: $y = \pm(b/a)x$. Slopes of the lines are
the square root of the number under the y^2 term over
the square root of the number under the x^2 term. The
shape of the curve depends on the location of the
minus sign, *not* the largeness of the numbers under the
x^2 or y^2 term.

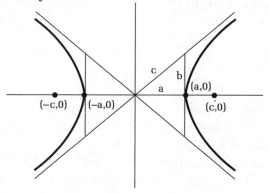

EXAMPLE 8—

Sketch and label $x^2/7 - y^2/11 = 1$.

Transverse vertices: $y = 0$ (set the letter after the minus sign equal to 0). $(\pm\sqrt{7},0)$. $c = \sqrt{(7 + 11)}$. Foci: $(\pm\sqrt{18},0)$. Asymptotes: $y = \pm(\sqrt{11}/\sqrt{7})x$.

NOTE

Curve does not hit y axis. If $x = 0$, $y = \pm(-11)^{1/2}$, which are imaginary.

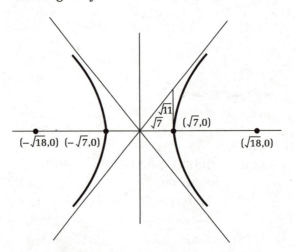

EXAMPLE 9—

Sketch and discuss $y^2/5 - x^2/9 = 1$.

Let $x = 0$. Transverse vertices are $(0,\pm\sqrt{5})$. $c = \sqrt{(5 + 9)}$. Foci are $(0,\pm\sqrt{14})$. Asymptotes: $y = \pm(\sqrt{5}/\sqrt{9})x$. The sketch is as shown here.

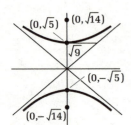

EXAMPLE 10—

Sketch and discuss $(y - 6)^2/5 - (x + 7)^2/9 = 1$.

This is the same as Example 9, except the "center" of the hyperbola, where the asymptotes cross, is no longer at (0,0). The center is (−7,6). $(x + 7)^2 = 0$; then $a = \sqrt{5}$, $\sqrt{5}$ above and below the center. $c = \sqrt{(5 + 9)}$. The foci

are $\sqrt{14}$ above and below the center. V($-7,6 \pm \sqrt{5}$).
F($-7,6 \pm \sqrt{14}$). Asymptotes: $y - 6 = \pm(\sqrt{5}/\sqrt{9})(x + 7)$.

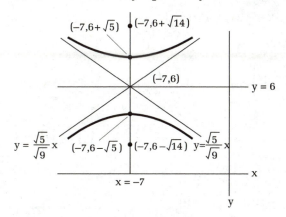

EXAMPLE 10 REVISITED

$$\frac{(y - 6)^2}{5} - \frac{(x + 7)^2}{9} = 1$$

Just like with the ellipse, let's clarify the location of
the vertices. In this case, the vertices occur north-south
of the center $(-7,6)$. So the x value of the vertices is the
same as the center, $(-7,6)$. $x = -7$. Substituting, we get

$$\frac{(y - 6)^2}{5} - \frac{(-7 + 7)^2}{9} = 1$$

$$(y - 6)^2 = 5$$

$$y = 6 \pm \sqrt{5}$$

and the vertices are $(-7,6 \pm \sqrt{5})$. For the same reason,
the foci are $(-7,6 \pm \sqrt{14})$. The asymptotes and the
sketch are the same!

EXAMPLE 11

Sketch and discuss $25x^2 - 4y^2 + 50x - 12y + 116 = 0$.

For the last time, we will complete the square, again a
little differently than the other times. We will use

exactly the same steps as for the ellipse, except for the minus sign.

$$25x^2 - 4y^2 + 50x - 12y + 116 = 0$$

$$25x^2 + 50x - 4y^2 - 12y = -116$$

$$25\left[x^2 + 2x + \left(\frac{2}{2}\right)^2\right] - 4\left[y^2 + 3y + \left(\frac{3}{2}\right)^2\right]$$

$$= -116 + 25\left(\frac{2}{2}\right)^2 - 4\left(\frac{3}{2}\right)^2$$

$$\frac{25(x+1)^2}{-100} - \frac{4\left(y+\frac{3}{2}\right)^2}{-100} = \frac{-100}{-100}$$

$$\frac{\left(y+\frac{3}{2}\right)^2}{25} - \frac{(x+1)^2}{4} = 1$$

The center is $(-1,-3/2)$. $V(-1,-3/2 \pm \sqrt{25})$. $F(-1,-3/2 \pm \sqrt{29})$. Asymptotes: $y + 3/2 = \pm(\sqrt{25}/\sqrt{4})(x + 1)$.

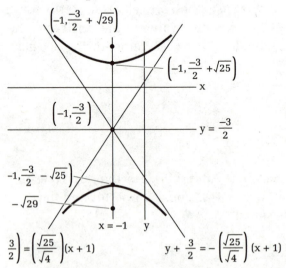

$$\frac{3}{2}\right) = \left(\frac{\sqrt{25}}{\sqrt{4}}\right)(x+1) \qquad y + \frac{3}{2} = -\left(\frac{\sqrt{25}}{\sqrt{4}}\right)(x+1)$$

Sometimes we have a puzzle. Given some information, can we find the equation? You must always draw the picture and relate the picture to its equation.

EXAMPLE 12—

Find the equation of the parabola with focus (1,3), directrix x = 11.

Drawing F and the directrix, the picture must be the one shown here. The vertex is halfway between the x numbers. So x = (11 + 1)/2 = 6. V(6,3). c = the distance between V and F = 5. The equation is $(y - 3)^2 = -4c(x - 6) = -20(x - 6)$. Remember, the minus sign is from the shape and c is always positive for these problems.

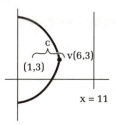

EXAMPLE 13—

Vertices are (2,3) and (12,3), and there is one focus (11,3). Find the equation of the ellipse.

The two vertices give the center [(12 + 2)/2,3] = (7,3). F(11,3). $(x - 7)^2/a^2 + (y - 3)^2/b^2 = 1$. a = 12 − 7 = 5. c = 11 − 7 = 4. $a^2 - b^2 = c^2$. $5^2 - b^2 = 4^2$. $b^2 = 9$ (no need for b). $(x - 7)^2/25 + (y - 3)^2/9 = 1$.

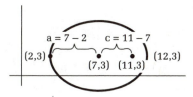

EXAMPLE 14—

Find the equation of the hyperbola with vertices (0,±6) and asymptotes y = ±(3/2)x.

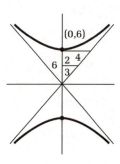

V(0,±6) says the center is (0,0) and the shape is $y^2/36 - x^2/a^2 = 1$. The slope of the asymptotes is 3/2 = square root of the number under y^2 over the square root of the number under the x^2 term. So 3/2 = 6/a. So a = 4. So $a^2 = 4^2 = 16$. The equation is $y^2/36 - x^2/16 = 1$.

This kind of question is shorter in length, but it does take practice. So practice!!!

CHAPTER 16

RIGHT AND NOT SO RIGHT-ANGLE TRIG, LAW OF SINES, LAW OF COSINES

Most geometry courses have a little right-angle trig. I am doing it the standard way, because the way I like to do it is too long for this book. If you want to see a better way, go to my *Precalc with Trig for the Clueless*. The trig part of the book is something you all will be able to understand. The rest depends on your algebra background.

We have a right triangle ABC, with the right angle at C. For angle A, AC is called the adjacent leg because it touches angle A. For angle A, BC is the opposite leg because it doesn't touch angle A. AB is of course the hypotenuse.

DEFINITION

$$\text{Sine A} = \sin A = \frac{BC}{AB} = \frac{\text{opposite}}{\text{hypotenuse}}$$

$$\text{Cosine A} = \cos A = \frac{AC}{AB} = \frac{\text{adjacent}}{\text{hypotenuse}}$$

$$\text{Tangent A} = \tan A = \frac{BC}{AC} = \frac{\text{opposite}}{\text{adjacent}}$$

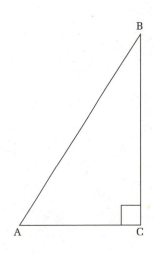

There are called trig ratios.

NOTE 1

If you extend or shrink the sides, if the angle stays the same, the ratio of the sides stay the same because of similar triangles. Sooo the sine, cosine, and tangent of the angle will always remain the same.

NOTE 2

To find these values, we either need tables or a calculator. We will use calculators because that is what calculators should be used for.

NOTE 3

For angle B, the opposite side for A is the adjacent side for B, and the adjacent side for A is the opposite side for B. The hypotenuse is always the hypotenuse.

NOTE 4

(There is a memory device to help remember these trig ratios, one that my father remembered from 1916 (anything lasting that long must be good). It is the Indian chief SOHCAHTOA: sine is opposite over hypotenuse; cosine is adjacent over hypotenuse; and tangent is opposite over adjacent. By the way there are Indian powwows in my area. I've asked them if they object to the term Indian. None of them did. Only politically correct people seem to want Native Americans.

Let's do some examples.

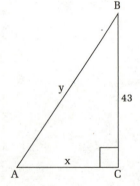

EXAMPLE 1

BC is 43 feet. Angle A is 73 degrees. Find all the missing parts.

1. Angle B = 90 − 73 = 17 degrees.

2. Tan 73° = opp/adj = 43/x. Sooo x = 43/tan 73°. By calculator x = 13.1 feet.

3. Sin 73° = opp/hyp = 43/y. y = 43/sin73°. By calculator y = 45.0 feet. Pretty easy huh?!!! Let's try
another.

EXAMPLE 2—

Find x, angles A and B.

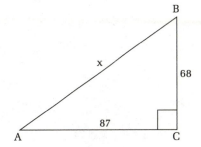

1. To get x, use old Pythagoras x = $\sqrt{87^2 + 68^2}$. By
calculator x = 110.4 feet.

2. Tan A = opp/adj = 68/87. To find the angle, you
must use tan^{-1} 68/85 = 38.7° = (m)∠A.
(m)∠B = 90° − 38.7° = 51.3°.

NOTE

To know what tan^{-1}, "arctan," or "inverse tan" is, you
must study more trig. It really isn't hard or shouldn't
be. At this point, all you need to know is that it gives
you the angle.

Some books do problems on the angle of elevation and
the angle of depression.

IMPORTANT NOTE

Make sure your calculator gives you degrees when you are doing angles.

DEFINITION

Angle of elevation: The angle you look up at. The bottom line is the horizon.

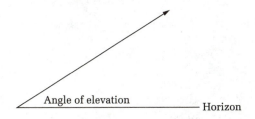

Angle of elevation   Horizon

DEFINITION

Angle of depression: The angle you look down at something. The top line is the horizon.

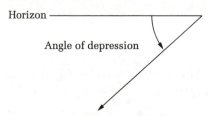

Horizon

Angle of depression

EXAMPLE 3

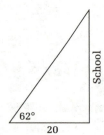

Sandy wants to measure the height of the school. If the angle of elevation from the ground is 62 degrees, and Sandy is 20 feet from the school, how tall is the school?

$$\tan 62° = \frac{\text{opp}}{\text{adj}} = \frac{x}{20} \qquad x = 20 \tan 62° = 37.6 \text{ feet tall!!}$$

You might actually try to measure the height of your school.

EXAMPLE 4

A man is looking down from the top of a 120-foot lighthouse. The angle of depression to a boat is 20°. How far away is the boat?

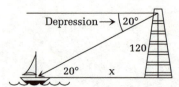

Depression → 20°

120

20°   x

Try to get the variable in the numerator. cot 20° = x/120. So x = 120 cot 20°. By calculator, the answer is 330 feet.

EXAMPLE 5

John is 200 feet from a building on which is located a tall antenna. The angle of elevation to the bottom of the antenna is 70°. The angle of elevation to the top of the antenna is 78°. How tall is the antenna?

There are two unknowns: the height of the building and the height of the antenna. We always want to try to get the unknown we don't want (the building) in the numerator. It makes the problem easier. In this problem, both unknowns are in the numerator when you use tangent.

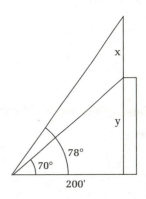

$$\tan 78° = \frac{(x + y)}{200} \qquad x + y = 200 \tan 78°$$

$$\tan 70° = \frac{y}{200} \qquad y = 200 \tan 70°$$

Subtract; factor.

$$x = 200(\tan 78° - \tan 70°) = 394 \text{ feet (big!)}$$

THE LAW OF SINES

When we don't have a right angle, we must use the law of sines or the law of cosines.

DEFINITION
Law of sines—a/sin A = b/sin B = c/sin C.

NOTE
For this exercise we will say sin 45° = 0.7, which is approximately correct.

We use the law of sines whenever we have any two angles and a side or two sides and an angle opposite one of those sides.

NOTE

Side is small letter; angle opposite is the same letter, only capitalized.

Two Angles and a Side

If we have two angles, we have the third angle. With a side, we learned, in geometry, that triangles are congruent by angle, side, angle. Therefore one solution is possible.

EXAMPLE 6—

A = 56°, B = 73°, and a = 20. Find all the other parts.

C = 180° − (56° + 73°) = 51°   a/sin A = b/sin B
20/sin 56° = b/sin 73°   b = 20 sin 73°/sin 56°   b = 263
a/sin A = c/sin C   20/sin 56° = c/sin 51°   c = 20 sin
51°/sin 56°   c = 19

Two Sides and an Angle Opposite One of Those Sides

If you draw the triangle, we have side-side-angle. From geometry, we know triangles are *not* congruent. This is called the ambiguous case. No triangle, one triangle, or two triangles are possible. Let us give one example of each.

EXAMPLE 7—

Let a = 10, A = 30°, b = 50. a/sin A = b/sin B. 10/
sin 30 = 50/sin B. sin B = (50 sin 30)/10 = 2.5. No triangle, since the sine is never bigger than 1.

EXAMPLE 8—

A = 135°, a = 70, b = 50. a/sin A = b/sin B. 70/
sin 135° = 50/sin B. sin B = (50 sin 135)/70. sin B = ½.
B = 30° or 180° − 30° = 150°.

135° + 30° = 165° is OK. 135° + 150° = 285° is no good—the sum of angles of a triangle is 180°. One triangle is possible. A = 135°, B = 30°, so C = 15°. a = 70, b = 50. a/sin A = c/sin C. 70/sin 135° = c/sin 15°. c = 70 sin 15°/sin 135°. c = 26.

EXAMPLE 9—

C = 30°. c = 5. d = 7. c/sin C = d/sin D. 5/sin 30 = 7/
sin D. sin D = (7 sin 30°)/5 = 0.7. D = 45° or D′ = 180 −
45 = 135°. 30 + 45, OK. 30 + 135, OK. Two triangles!!!

Triangle 1: C = 30°, D = 45°, E = 105°. c = 5, d = 7,
e/sin 105° = 5/sin 30°. e = 9.7.

Triangle 2: C = 30°, D′ = 135°, E′ = 15°. c = 5, d = 7,
e′/sin 15° = 5/sin 30°. e′ = 2.6.

EXAMPLE 10—

The angle between Zeb and Sam as seen by Don is 70°.
The angle between Don and Zeb as seen by Sam is 62°.
Zeb and Sam are 70 feet apart. How far apart are Don
and Sam?

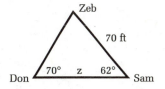

   Angle at Zeb is 48°. z/sin 48° = 70/sin 70°. z = 55
feet.

THE LAW OF COSINES

We use the law of cosines whenever we have three
sides or two sides and the included angle.

DEFINITION

Law of cosines—$c^2 = a^2 + b^2 - 2ab \cos C$. Again, note
angle C opposite side c. In each case only one triangle
is possible, since triangles are congruent by side-side-
side or side-angle-side.

EXAMPLE 11—

a = 3, b = 5, c = 7. $c^2 = a^2 + b^2 - 2ab \cos C$. $7^2 = 3^2 + 5^2 -$
2(3)(5) cos C. Careful of the arithmetic. 15 = −30 cos C.
cos C = −½. C = 120° (quadrant II).

NOTE

Whenever you have four parts of a triangle, it is easier
to use the law of sines. 3/sin A = 7/sin 120°. A = 22°.
B = 180° − (22 + 120) = 38°.

EXAMPLE 12—

$x = 10$, $y = 20$, $Z = 40°$. $z^2 = x^2 + y^2 - 2xy \cos Z$. $z^2 = 10^2 + 20^2 - 2(10)(20) \cos 40°$. $z = 13.9$. $13.9/\sin 40° = 10/\sin X$. $\sin X = 10(\sin 40°)/13.9$. $X = \sin^{-1} (10 \sin 40°/13.9)$. $X = 28°$. $Y = 180° - (28° + 40°) = 112°$.

EXAMPLE 13—

A plane travels east for 200 miles. It turns at a 25° angle to the north for 130 miles. How many miles from home is the plane?

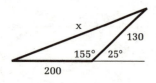

$x^2 = 200^2 + 130^2 - 2(200)(130) \cos 155°$

$x = 323$ miles

That's all the trig we need for now. More? See *Precalc with Trig for the Clueless.*

CHAPTER 17

MISCELLANEOUS: LOCUS, PARALLEL LINES, LARGER AND SMALLER SIDES AND ANGLES

The following topics do not fit into a particular section easy. So let's do them here.

LOCUS

My heart feels bad because these problems are rarely done any more. But let's do a few of these here.

LOCUS

A point or the set of all points and only those points satisfying a condition or conditions. Some examples will explain this. You should draw the picture if possible.

EXAMPLE 1—

The set of all points at a distance r from a given point P.

1 dimension: Draw the line through P (that is the one dimension): locus is 2 points, distance away from P.

2 dimensions (x y plane) locus: circle, radius r, center P.

3 dimensions (space) locus: sphere, radius r, center P.

EXAMPLE 2—

The set of all points equidistant from 2 points, A and B.

1 dimension: Draw the line segment AB. The midpoint.

2 dimensions: The perpendicular bisector of AB.

3 dimensions: Draw AB. Draw the perpendicular bisector of AB. Draw the plane through AB and its perpendicular bisector (remember, there is only one). Locus: a plane through the perpendicular bisector and perpendicular to the plane you just drew. (3-D can be hard to visualize.)

EXAMPLE 3—

The set of all points at a distance r from line "a" parallel to line a.

1 dimension: Impossible since the line is the one dimension.

2 dimensions: Two lines parallel to line "a" at a distance r from the line.

3 dimensions: A cylinder with line "a" through its center, radius of the circular base r.

These are wonderful visualization problems and fun!! I hope the books return them to your course.

PARALLEL LINES

Theorem

If three or more parallel lines are cut by two transversals, the lines make proportional segments on each transversal.

EXAMPLE 4—

The proportion is x/5 = 7/8.

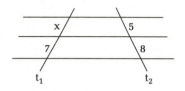

$8x = 35$ $x = 35/8.$

They can be made a little harder, but not much.
  I knew I forgot something!

LARGER AND SMALLER SIDES AND ANGLES

Theorem

In a triangle, the largest side lies opposite the largest angle and the largest angle lies opposite the largest side.

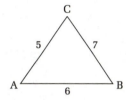

EXAMPLE 5—

The largest angle is A because it is opposite the longest side 7. Next is angle C. The smallest is angle B.

EXAMPLE 6—

Which is the largest side? In △ABC, the largest side is BC since it is opposite the largest angle, 62 degrees. In △BCD, the largest side is BD since it lies opposite the 61 degree angle in the second triangle.

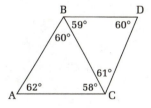

  Buuuuut BC and BD are in the same triangle!!! Since BD is the largest side in this triangle, BD is the largest side.

  Actually this section is slightly out of place since in trig the law of sines tells you exactly the relationship between angles and sides. But most of you will not do the law of sines. If you are curious, just look.

  At this point I apologize since there are several topics that I would love to include but I am not because they have either totally disappeared from books or almost. First there are more similar triangle proofs.

Second there are inequality proofs. And finally, the topic that is the most fun and is good geometry also, is doing construction problems with ruler and compass. Such examples are bisecting a line segment, bisecting an angle, constructing a perpendicular bisector from a point to a line, drawing a hexagon, drawing a pentagon, etc. Someday when these topics return, I'm sure McGraw-Hill will allow me to write a new edition.

Until then, let me finish up with a little quiz for you, an always-sometimes-never quiz.

CHAPTER 18

ALWAYS-SOMETIMES-NEVER QUESTIONS

To help you find out how well you know geometry and because teachers very often ask these questions, we will do always-sometimes-never questions. You must fill in one of these choice in the blank space in the questions.

Always means always true.

Sometimes means at least true one time and at least false one time.

Never means always false.

Let's go!!!!!

INTRODUCTORY STUFF

1. In a plane, a series of connected line segments is _____ a polygon.

2. A bisector of a line segment _____ passes through the midpoint of the line segment.

3. When two lines intersect, all the vertical angles are _____ equal (congruent).

4. The complement of an acute angle is _____ acute.

5. The supplement of an acute angle is _____ acute.

6. If two angles are supplementary, they are _____ equal (congruent).

7. In an isosceles triangle, base angles are _____ obtuse.

8. If WX is the perpendicular bisector of YZ, then YZ _____ is the perpendicular bisector of WX.

9. Two angles with a common vertex are _____ adjacent angles.

10. The bisectors of two adjacent supplementary angles are _____ perpendicular.

11. If equals are multiplied by equals their products are _____ equal.

12. If two lines meet, their vertical angles are _____ supplementary.

13. A theorem is a statement that is _____ assumed to be true.

14. Two complementary angles are _____ equal (congruent).

CONGRUENT AND OTHER TRIANGLES

15. Two triangles are _____ congruent if three angles of one are equal (congruent) to three angles of another.

16. Two triangles are _____ congruent if two sides and an angle of one =($\cong$) to two sides and an angle of another.

17. Two triangles are _____ congruent if two angles and a side of one =(≅) to two angles and a side of another.

18. Two triangles are _____ congruent if three sides of one are equal (congruent) to three sides of another.

19. Two right triangles are _____ congruent if the hypotenuse and one leg of one is equal (congruent) to the same in another.

20. Two right triangles are _____ if one leg and one acute angle =(≅) to the same in another.

21. Two isosceles triangles are _____ congruent if the vertex and one base angle of one are =(≅) to the same in the other.

22. Two isosceles triangles are _____ congruent if one leg and the vertex angle of one =(≅) the same in another.

23. In an isosceles triangle two medians are _____ equal (congruent).

24. The altitude to the base of an isosceles triangle _____ bisects the vertex angle.

25. The altitude to the base of an isosceles triangle _____ bisects the base.

26. An equilateral triangle is _____ equiangular.

QUADS, PARALLELS, PERPS, AND ANGLE SUMS

27. Two lines cut by a transversal _____ have alternate interior angles equal (congruent).

28. Two parallel lines cut by a transversal _____ have interior angles on the same side of the transversal =(≅).

29. The bisector of interior angles on the same side of the transversal of parallel lines are _____ =(≅).

30. The bisector of interior angles on the same side of the transversal of parallel lines are _____ ⊥.

31. Alternate exterior angles of two parallel lines are _____ supplementary.

32. If a line is parallel to one of two parallel lines it is _____ parallel to the other.

33. If a line is parallel to one of two perpendicular lines, it is _____ perpendicular to the other.

34. (The measure of) an exterior angle of a polygon _____ 80 degrees.

35. If the diagonals of a quadrilateral are equal (congruent), the figure is _____ a rhombus.

36. If the diagonals are equal (congruent), the quadrilateral is _____ a rectangle.

37. If the diagonals of a quadrilateral bisect the angles of the figure, the figure _____ is a rhombus.

38. If the diagonals of a quadrilateral form four congruent triangles, it is _____ a rhombus.

39. A quadrilateral with four right angles is _____ a square.

40. The bisector of the opposite angles of a parallelogram _____ coincide.

41. The sum of the (measures of the) interior angles of a quadrilateral is _____ the same as the sum of the (measures of the) exterior angles.

42. The diagonals of a trapezoid are _____ equal (congruent).

43. If two sides of a quadrilateral are equal (congruent) and two are parallel, the figure is _____ a parallelogram.

CIRCLES

44. A chord is _____ a diameter.

45. A chord is _____ a radius.

46. A diameter is _____ a chord.

47. An angle formed by two tangents _____ equals in degrees to half the (measure of the) intercepted arc.

48. If two arcs of a circle are the equal (congruent), their chords are _____ equal (congruent).

49. The perpendicular bisector of a chord _____ passes through the center of the circle.

50. A triangle inscribed in a semicircle is _____ a right angle.

51. A parallelogram inscribed in a circle is _____ two acute angles.

52. If two angles intercept the same arc, they are _____ equal (congruent).

53. If a chord in one circle is equal (congruent) to a chord in another circle, the minor arcs they intercept are _____ equal (congruent).

54. If two circles are concentric (having the same center), any chords of the larger circle, which are also tangent to the smaller circle, are _____ equal (congruent).

55. Any angle inscribed in an arc less than a semicircle is _____ obtuse.

SIMILAR TRIANGLES, SEGMENTS, AND A LITTLE ON TRIG

56. Congruent polygons are _____ similar.

57. A median of a triangle _____ divides the triangle into two similar triangles.

58. The sine of an angle _____ is the cosine of its complement.

59. Similar triangles are _____ congruent triangles.

60. Two isosceles triangles are _____ similar if their vertex angles are equal (congruent).

61. Two right triangles are _____ similar if one acute angle of one is equal (congruent) to another.

62. As an angle increases, the tangent of that angle _____ increases.

63. As an angle increases, the cosine of that angle _____ increases.

64. A rhombus whose side is 10 inches _____ has a diagonal of 20 inches.

AREAS AND VOLUMES

65. If the ratio of the area of two similar figures is 4/9, the ratio of their volumes is _____ 8/27.

66. If two polygons are congruent, they are _____ equal in area.

67. Triangles with the same base and same altitude to that base are _____ congruent.

68. If the ratio of two similar triangles is 1/7, the larger base is _____ 7 times the smaller base.

69. The area of a rhombus is _____ one-half the product of its diagonals.

70. If the sides of two rhombuses are equal (congruent), the areas are _____ equal.

71. If the volumes of two cylinders are equal, the cylinders are _____ congruent.

72. If the radii of two spheres are the same, their volumes are _____ equal.

73. If two cylinders have the radius of the base the same and the height the same, they are _____ congruent.

74. The ratio of their area of two unequal circles is _____ the same as the ratio of their radius.

75. If the bases of a triangle are the same, the ratio of their areas are _____ the ratio of the altitudes drawn to that base.

   This is just a sampling. Let's take a look at the answers.

1. Sometimes. Only when it is closed.

2. Always.

3. Sometimes. When they are perpendicular.

4. Always.

5. Never. Always obtuse.

6. Sometimes. Again when they are right angles.

7. Never. The sum of the angles must be 180; two obtuse angles would total more than 180.

8. Sometimes. Allllways perpendicular buut sometimes bisect.

9. Sometimes. When there is a common side between.

10. Always.

11. Always. An axiom.

12. Sometimes. When they are perpendicular.

13. Never. Theorems always must be proven, even if we don't do them; postulates and axioms are assumed to be true.

14. Sometimes. When they are each 45 degrees.

15. Sometimes. Always similar but can be bigger or smaller.

16. Sometimes. Only if the angle is between the sides.

17. Always. Two angles mean three angles and we allllways have ASA.

18. Always.

19. Always. HL.

20. Always. ASA.

21. Sometimes. The same angles mean similar, but not necessarily congruents, since at least one pair of congruent sides is necessary.

22. Always. SAS.

23. Always. Medians from each base angle.

24. Always.

25. Always.

26. Always.

27. Sometimes. They must be parallel!!!!!! These questions can be verrrrry tricky.

28, 29. Sometimes. When the transversal is perpendicular to them.

30. Always!! Draw the picture.

31. Sometimes. See 28 and 29.

32. Always.

33. Sometimes. Always if they are all in one plane.

34. Never. 360 is not divisible by 80.

35. Sometimes. If it is also a square.

36. Sometimes. It might be an arrowhead or an isosceles trapezoid.

37. Always.

38. Always.

39. Sometimes. It might only be a rectangle.

40. Sometimes. When it is also a rhombus or a square.

41. Always.

42. Sometimes. When it is also isosceles.

43. Sometimes. It might be an isosceles trapezoid.

44. Sometimes.

45. Never.

46. Always. The biggest chord.

47. Sometimes; really tough question.

48. Always.

49. Always. Draw the picture.

50. Never. An angle inscribed in a semicircle is a right angle.

51. Never. It must always be at least a rectangle. Again draw it.

52. Sometimes.

53. Always.

54. Always.

55. Never. It is acute since half of less than 180 is less than 90 degrees.

56. Always.

57. Sometimes.

58. Always.

59. Sometimes.

60. Always.

61. Always.

62. Always.

63. Never.

64. Never.

65. Always. Ratio of similarity is the square root, 2/3; volumes would be 2/3 cubed or 8/27.

66. Always. Congruent means the same in every way!!!!!

67. Sometimes. Right triangles and not right triangles might have the same area.

68. Always.

69. Always.

70. Sometimes. The angles at their bases could be different; so are their heights and diagonals.

71. Sometimes.

72. Always.

73. Sometimes!!!!! Cylinder can be at an angle!!!

74. Never. Ratio of areas is the square of the ratio of radii (the ratio of similarity).

75. Always.

As you can tell, some of these are easy and some are not so easy, but all of them make you think and know your geometry. Because they make you think and because they are short answers are the reasons teachers ask these questions. Perhaps your book has a few more for you to practice.

A RADICAL CHAPTER

When I started teaching, I thought that this topic caused students many problems. Maybe it was me, and maybe I was wrong, but now my students learn this topic super well. I'm sure you will too.

DEFINITION

$\sqrt{a}$, "square root of a": the number b, such that (b)(b) = a.

$\sqrt{9}$ = 3, because (3)(3) = 9. $\sqrt{25}$ = 5 because (5)(5) = 25. 9 and 25 are called *perfect squares* because they have exact square roots.

$\sqrt[3]{a}$, "cube root of a": the number c, such that (c)(c)(c) = a. $\sqrt[3]{64}$ = 4 because (4)(4)(4) = 64.

$\sqrt[4]{a}$, "the fourth root of a": the number h, such that (h)(h)(h)(h) = a. $\sqrt[4]{81}$ = 3 because (3)(3)(3)(3) = 81.

$\sqrt[n]{a}$, "the nth root of a": the number b such that (b)(b)(b) . . . (b) = a
(n factors).

EXAMPLE 1—

$\sqrt{16} = 4$; $-\sqrt{16} = -4$ (minus sign is on the outside of square root).

$\pm\sqrt{16} = \pm 4$ ($\pm\sqrt{16}$ means two numbers, $+\sqrt{16}$ and $-\sqrt{16}$).

$\sqrt{-16}$ is not a real number because a positive times a positive is positive and a negative times a negative is positive. No real number squared gives a negative. Also, $\sqrt{0} = 0$. Buuuut . . .

EXAMPLE 2—

$\sqrt[3]{1000} = 10$; $\sqrt[3]{-125} = -5$; $\sqrt[3]{0} = 0$

$-\sqrt[3]{-27} = -(-3) = 3$; $\pm\sqrt[3]{216} = \pm 6$; $\pm\sqrt[3]{-8} = \mp 2$

(The chart is on the next page.)

As you do these problems, you should keep this list in front of you. Don't go home and memorize. But as you do the problems, you will be surprised how many you learn from repetition.

EXAMPLE 3—

Solve for x:

$x^2 - 16 = 0$

$x^2 = 16$

Two roots.

$x = \pm\sqrt{16} = \pm 4$

Powers

	Square	Cube	4th	5th	6th	7th	8th	9th	10th	11th	12th
2	4	8	16	32	64	128	256	512	1024	2048	4096
3	9	27	81	243	729						
4	16	64	256	1024							
5	25	125	625	3125							
6	36	216									
7	49	343									
8	64	512									
9	81	729									
10	100	1000									
11	121	1331									
12	144	1728									
13	169										
14	196										
15	225										
16	256										
17	289										
18	324										
19	361										
20	400										
21	441										
22	484										
23	529										
24	576										
25	625										
26	676										
27	729										
28	784										
29	841										
30	900										
31	961										
32	1024										

The chart is read backward and forward. 5 cubed is 125, buuuut also the cube *root* of 125 is 5.

NOTES

You will find that most of these numbers repeat over and over; some now and some later on.

Try not to use calculators. If you put these numbers in your brain, many things go easier for you.

Some numbers are not perfect squares. 2 is not a perfect square. There is no integer or rational number that gives you 2 when you square it. $\sqrt{2}$ is *irrational*.

An *irrational number* is any number that cannot be written as a terminating or repeating decimal (any number that is not rational).

π, $\sqrt{7}$, and $\sqrt[3]{5}$ are all examples of irrational numbers.

Remember, writing 3.14 or 22/7 are only approximations for π.

A *real* number is any decimal number.

The real numbers are the rationals plus the irrationals.

Now that we know what square and other roots are, we would like to simplify them; add, subtract, multiply, and divide them; and apply them.

95 percent of the time, the shortest way is the simplest. I was always taught to use the K.I.S.S. method of teaching—keep it simple, stupid. However, the radical unit is one of the few times we don't use the shortest method. The method is short, just not the shortest.

SIMPLIFYING RADICALS

EXAMPLE 1—

Simplify

$\sqrt{18}$

$= \sqrt{2\boxed{(3)(3)}}$

$= 3\sqrt{2}$

We write 18 as the product of primes: 18 = 2(3)(3).

Two 3s circled on the inside become one 3 on the outside. Why? Because $\sqrt{(3)(3)} = (\sqrt{9}) = 3$.

EXAMPLE 2—

Simplify

$\sqrt{72}$

$= \sqrt{\boxed{(2)(2)}(2)\boxed{(3)(3)}} = 2 \times 3\sqrt{2} = 6\sqrt{2}$

Break 72 into primes. No matter how you do it, you will get the same ones.

EXAMPLE 3—

Simplify

$10\sqrt{125}$

$= 10\sqrt{\boxed{(5)(5)}(5)} = 10(5)\sqrt{5} = 50\sqrt{5}$

EXAMPLE 4—

Simplify

$\sqrt{27b^7} = \sqrt{(3)\boxed{(3)\ (3)}\ \boxed{(b)(b)}\ \boxed{(b)(b)}\ (b)(b)\ (b)} = 3b^3\sqrt{3b}$

EXAMPLE 5—

Simplify

$\sqrt{a^{11}} = \sqrt{\boxed{aa}\boxed{aa}\boxed{aa}\boxed{aa}\boxed{aa}\,a} = a^5\sqrt{a}$

Short way!!!! Nice! $\sqrt{}$ means $\sqrt[2]{}$. Divide 2 into 11. We get 5 groups of 2. So a^5 goes on the outside. The remainder is 1. a goes on the inside. That's all!

EXAMPLE 6—

Simplify

$$4y^3\sqrt{25y^3} = (4y^3)(5y)\sqrt{y} = 20y^4\sqrt{y}$$

EXAMPLE 7—

Simplify

$$\sqrt[3]{72a^{14}b^9}$$

With cube roots, we circle three of the same factors. (Fourth roots we circle 4 of the same.) 3 into 14 is 4 with a remainder of 2. 3 into 9 is 3 with no remainder. Soooooo,

$$\sqrt[3]{72a^{14}b^9} = 2a^4b^3\sqrt[3]{(3)(3)a^2} = 2a^4b^3\sqrt[3]{9a^2}$$

Since

$$\sqrt[3]{72} = \sqrt[3]{\boxed{(2)\,(2)\,(2)}(3)(3)} = 2\sqrt[3]{(3)(3)}$$

Not too bad!!

ADDING AND SUBTRACTING RADICALS

Adding and subtracting radicals is the same as adding and subtracting like terms. With like radicals, you add or subtract the coefficients. Unlike radicals cannot be combined.

EXAMPLE 1—

Simplify

$$4\sqrt{2} - 5\sqrt{7} - 7\sqrt{2} - 10\sqrt{7}$$

The answer issss

$$-3\sqrt{2} - 15\sqrt{7}$$

EXAMPLE 2—

Simplify

$3\sqrt{27} + 5\sqrt{8} + 6\sqrt{12}$

We must simplify each radical first.

$= 3\sqrt{(3)(3)(3)} + 5\sqrt{(2)(2)(2)} + 6\sqrt{(2)(2)(3)}$

$= (3)(3)\sqrt{3} + 5(2)\sqrt{2} + (6)(2)\sqrt{3} = 9\sqrt{3} + 10\sqrt{2} + 12\sqrt{3}$

$= 21\sqrt{3} + 10\sqrt{2}$

EXAMPLE 3—

Simplify

$\sqrt[3]{128y^8} + 10y\sqrt[3]{54y^5}$

$= \sqrt[3]{(2)(2)(2)(2)(2)(2)(2)(y)(y)(y)(y)(y)(y)(y)(y)}$

$+ 10y\sqrt[3]{(2)(3)(3)(3)(y)(y)(y)(y)(y)}$

$= 4y^2\sqrt[3]{2y^2} + 10y(3y)\sqrt[3]{2y^2} = 34y^2\sqrt[3]{2y^2}$

MULTIPLYING RADICALS

There are two types. Sometimes they occur together.

If a, b $\geq$ 0, then $\sqrt{a} \times \sqrt{b} = \sqrt{ab}$. In fact, if n is positive even, $\sqrt[n]{a} \times \sqrt[n]{b} = \sqrt[n]{ab}$. If n is odd positive, then a and b can be any real number and $\sqrt[n]{a} \times \sqrt[n]{b} = \sqrt[n]{ab}$. In particular, $\sqrt[3]{a} \times \sqrt[3]{b} = \sqrt[3]{ab}$.

Also, $a\sqrt{b} \times c\sqrt{d} = ac\sqrt{bd}$. You multiply the insides and multiply the outsides. Again, both b and d must be greater than or equal to 0.

EXAMPLE 1—

Multiply

$3\sqrt{6} \times 5\sqrt{7}$

The answer is $15\sqrt{42}$.

EXAMPLE 2—

Multiply (the easier way)

$7\sqrt{8} \times 10\sqrt{6}$

$= 70\sqrt{(2)(2)\,(2)(2)\,(3)}$

$= (70)(2)(2)\sqrt{3} = 280\sqrt{3}.$

If you first multiplied 6 times 8, the first thing you would have to do is break it down again. So write each one in terms of primes first to avoid wasting effort.

EXAMPLE 3—

$4\sqrt{2}(3\sqrt{5} + \sqrt{7} - 5 + 8\sqrt{6} + 3\sqrt{2})$

$= 12\sqrt{10} + 4\sqrt{14} - 20\sqrt{2} + 32\sqrt{12} + 12\sqrt{4}$

Multiply outsides Multiply insides	Multiply insides only	Multiply outsides only	Multiply outsides Multiply insides	Multiply outsides Multiply insides
			$\sqrt{12} = 2\sqrt{3}$	$\sqrt{4} = 2$

$= 12\sqrt{10} + 4\sqrt{14} - 20\sqrt{2} + 64\sqrt{3} + 24$

If you couldn't combine inside the parentheses before multiplication, you won't be able to combine terms after the multiplication.

EXAMPLE 4—

Multiply

$(2\sqrt{3} + 4\sqrt{2})(6\sqrt{3} + 2\sqrt{2})$

This looks familiar. Aha!! We must *foil*:

$(2\sqrt{3} + 4\sqrt{2})(6\sqrt{3} + 2\sqrt{2})$

$= 12\sqrt{9} + 4\sqrt{6} + 24\sqrt{6} + 8\sqrt{4}$

$= 36 + 28\sqrt{6} + 16 = 52 + 28\sqrt{6}$

Or, to be fancy,

$4(13 + 7\sqrt{6})$

DIVIDING RADICALS

The last operation is division, but in order to fully do this, we have to rationalize the denominator. Soooo, division will be in the middle of two sections on rationalizing.

Rationalize the Denominator I

Suppose we have $1/\sqrt{2}$. In the past the reason we never wanted radicals in the bottom was the problem of finding an approximate decimal value for $1/\sqrt{2}$. $\sqrt{2}$ is approximately 1.414 (which you should know, as you should $\sqrt{3}$, which is the year George Washington was born—1.732). To find the value of $1/\sqrt{2} = 1/1.414$, we had to divide $1.414\overline{)1.000000}$. This had two great disadvantages: this is very long division, and it is inaccurate because the divisor, 1.414, is rounded off. Now suppose we rationalize the denominator. Look what happens. . . .

EXAMPLE I—

Rationalize

$$\frac{1}{\sqrt{2}} = \frac{1}{\sqrt{2}} \times \frac{\sqrt{2}}{\sqrt{2}} = \frac{\sqrt{2}}{2}$$

By multiplying the top and bottom by $\sqrt{2}$, the arithmetic is much easier to do because . . . if you take $\sqrt{2}/2 = 1.414/2 = 2\overline{)1.414}$, first you are dividing by 2 (very easy) and second, when you round off the top, it is more accurate. However, we really don't have to do this with calculators. In math we almost always have to rationalize the numerator. *Math is behind the times.* I have found that when math is behind the times, math teaching is good. Unfortunately, as I write this book, math is the most progressive subject. In my view, math teaching today is at its worst and so is math learning. This is why I wrote this book!!!! Enough talk. I talk too much. Let's go on!!

EXAMPLE 2—

Rationalize the denominator

$$\frac{7}{\sqrt{75}}$$

Simplify the bottom first!!!

$$\sqrt{75} = \sqrt{3(5)(5)} = 5\sqrt{3}$$

Soooo,

$$\frac{7}{\sqrt{75}} = \frac{7}{5\sqrt{3}} = \frac{7}{5\sqrt{3}} \times \frac{\sqrt{3}}{\sqrt{3}} = \frac{7\sqrt{3}}{15}$$

EXAMPLE 3—

Rationalize the denominator

$$\frac{9}{\sqrt{8a^5}} = \frac{9}{2a^2\sqrt{2a}} = \frac{9}{2a^2\sqrt{2a}} \frac{\sqrt{2a}}{\sqrt{2a}} = \frac{9\sqrt{2a}}{4a^3}$$

Not too bad. Let's do division!!!

Division

Division is based on one simple principle:

$$\sqrt{\frac{a}{b}} = \frac{\sqrt{a}}{\sqrt{b}}$$

where a is not negative, buuuut b must be positive!

Let's do a bunch of problems!!!

EXAMPLE 1—

$$\sqrt{\frac{4}{9}} = \frac{\sqrt{4}}{\sqrt{9}} = \frac{2}{3}$$

Really easy.

EXAMPLE 2—

$$\sqrt{\frac{3}{25}} = \frac{\sqrt{3}}{\sqrt{25}} = \frac{\sqrt{3}}{5}$$

EXAMPLE 3—

$$\sqrt{\frac{7}{27}} = \frac{\sqrt{7}}{\sqrt{27}} = \frac{\sqrt{7} \times \sqrt{3}}{3\sqrt{3} \times \sqrt{3}} = \frac{\sqrt{21}}{9}$$

Notice that once you divide, the problems are just like those in the first part of this section.

EXAMPLE 4—

$$\sqrt{\frac{2a^3b^7}{6a^{10}b^2}} = \sqrt{\frac{1b^5}{3a^7}} = \frac{\sqrt{b^5}}{\sqrt{3a^7}} = \frac{b^2\sqrt{b}}{a^3\sqrt{3a}} \times \frac{\sqrt{3a}}{\sqrt{3a}} = \frac{b^2\sqrt{3ab}}{3a^4}$$

Pretty long, but not bad either. Let's try some more.

EXAMPLE 5—

$$\frac{\sqrt{10a^4}}{\sqrt{35a^7}} = \sqrt{\frac{10a^4}{35a^7}} = \sqrt{\frac{2}{7a^3}} = \frac{\sqrt{2}}{\sqrt{7a^3}} = \frac{\sqrt{2} \times \sqrt{7a}}{a\sqrt{7a} \times \sqrt{7a}}$$

$$= \frac{\sqrt{14a}}{7a^2}$$

We make one square root, simplify, and then break it up.

Last, let's do one with addition.

EXAMPLE 6—

$$\sqrt{10a} + \frac{1}{\sqrt{10a}} = \frac{\sqrt{10a}}{1} + \frac{1 \times \sqrt{10a}}{\sqrt{10a} \times \sqrt{10a}}$$

$$= \frac{\sqrt{10a}}{1} + \frac{\sqrt{10a}}{10a}$$

$$= \frac{10a \times \sqrt{10a}}{10a \times 1} + \frac{1\sqrt{10a}}{10a} = \frac{(10a + 1)\sqrt{10a}}{10a}$$

Rationalize the Denominator II

In the previous parts of this section, when we rationalized the denominator, the bottom was a *monomial* (one term). Let's see what happens when the bottom is a *binomial*.

EXAMPLE 1—

Rationalize the denominator

$$\frac{2\sqrt{3}}{4\sqrt{2} + \sqrt{7}}$$

To rationalize, we multiply top and bottom by the *conjugate* of the bottom, $4\sqrt{2} - \sqrt{7}$. To get the conjugate, one changes the sign in the middle.

$$\frac{2\sqrt{3} \times (4\sqrt{2} - \sqrt{7})}{(4\sqrt{2} + \sqrt{7}) \times (4\sqrt{2} - \sqrt{7})} = \frac{8\sqrt{6} - 2\sqrt{21}}{25}$$

Multiply out the top, *foil* the bottom and simplify if possible. The bottom will always be an integer.

EXAMPLE 2—

Rationalize the denominator

$$\frac{\sqrt{c} \times (\sqrt{c} + \sqrt{d})}{(\sqrt{c} - \sqrt{d}) \times (\sqrt{c} + \sqrt{d})} = \frac{c + \sqrt{cd}}{\sqrt{c} - \sqrt{d}}$$

Note that both terms don't have to be square roots. The conjugate of $6 + \sqrt{x}$ is $6 - \sqrt{x}$. The conjugate of $\sqrt{y} + 3$ is $\sqrt{y} - 3$.

That's all for now. If you've finished reading this whole book and need some algebra review, then read *Algebra for the Clueless*. Algebra, OK? Then you might want *Precalc with Trig for the Clueless*.

Need to study math preparation for the SAT? Read *SAT Math for the Clueless*.

Good luck. I hope you are really starting to enjoy math.

INDEX

ACKNOWLEDGMENTS

I would like to thank my wife Marlene who makes life worth living, my wonderful children, Sheryl and Eric, their mates, Glenn and Wanda, and my grandchildren, Kira and Evan, who prove that although having children is great, grandchildren are the purest form of pleasure.

I would like to thank my brother Jerry, my parents Cele and Lee Miller, and my wife's parents Edith and Siebeth Egna.

I also thank Dr. Robert Urbanski of Middlesex County Community College, Martin Levine of Market Source, Bill Summers of CCNY, Bernice Rothstein of CCNY, Sy Solomon of MCCC, Hazel Spencer of Miami of Ohio, Libby Alam of CCNY, Efua Tonge of CCNY, and Sharon Nelson of Rutgers.

I would like to thank the present and past staff at McGraw-Hill: Barbara Gilson, Deborah Aaronson, Maureen Walker, John Carleo, John Aliano, David Beckwith, Pat Koch, Michele Bracci, and Mary Loebig Giles.

As usual the last thanks go to three people who helped me keep my spirits up when it appeared no one wanted the books that now number seven: a great friend Gary Pitkofsky, another terrific friend and fellow lecturer David Schwinger, and my sharer of dreams, my cousin Keith Robin Ellis.

ABOUT BOB MILLER . . . IN HIS OWN WORDS

I received my B.S. and M.S. in math from Brooklyn Poly, now Polytechnic University, after graduating from George W. Hewlett High School, Hewlett, New York. After my first class, which I taught as a substitute for a full professor, one student told another upon leaving the room that "at least now we have someone who can teach the stuff." I was forever hooked on teaching. Since Poly, I have taught at Westfield State College, Rutgers, and the last 32 years at the City College of New York. No matter how bad I feel before class, I always feel great after I start teaching. I am always delighted when a student tells me that he or she always hated math and could never learn it, but that taking a class with me has made math understandable and even enjoyable. I have a fantastic wife, Marlene; a wonderful daughter, Sheryl; a terrific son, Eric; and a great son-in-law, Glenn. The newest members of my family are my adorable, brilliant granddaughter, Kira Lynn, 3 years old at this writing, and my handsome brilliant grandson, Evan Ross, 11 months old at this writing. Annnd waiting in the wings is Wanda, the bride-to-be of my son. My hobbies are golf, bowling, bridge, and crossword puzzles. Someday I hope a publisher will allow me to publish

the ultimate calculus text and the ultimate high school text so that our country can remain number one forever.

To me, teaching math always is a great joy. I hope I can give some of this joy to you.